Getting New Things Done

Getting New Things Done

*Networks, Brokerage, and the
Assembly of Innovative Action*

David Obstfeld

Stanford Business Books
An Imprint of Stanford University Press
Stanford, California

Stanford University Press
Stanford, California

Special discounts for bulk quantities of Stanford Business Books are available to corporations, professional associations, and other organizations. For details and discount information, contact the special sales department of Stanford University Press. Tel: (650) 725-0820, Fax: (650) 725-3457

Printed in the United States of America on acid-free, archival-quality paper

Library of Congress Cataloging-in-Publication Data

Names: Obstfeld, David, 1958– author.
Title: Getting new things done : networks, brokerage, and the assembly of
    innovative action / David Obstfeld.
Description: Stanford, California : Stanford Business Books, an imprint of
    Stanford University Press, 2017. | Includes bibliographical references and
    index.
Identifiers: LCCN 2016051135| ISBN 9780804760508 (cloth : alk. paper) | ISBN
    9781503603097 (epub)
Subjects: LCSH: Business networks. | Social networks. | Social skills. |
    Project management. | Technological innovations—Management.
Classification: LCC HD69.S8 O27 2017 | DDC 658.4036—dc23
LC record available at https://lccn.loc.gov/2016051135

Typeset by Classic Typography in 10.5/14 Bembo

Contents

Acknowledgments

Writing this book would simply not have been possible without the help of friends, colleagues, and family. Some time ago I was persuaded that a book-length exposition of my ideas on networks, knowledge, and creative projects was necessary to convey an important, untold story about innovative action.

I am first and foremost grateful to the men and women of NewCar who made this story possible. They welcomed me, allowed me to observe them at work, and generously shared their expertise and experiences. I especially want to acknowledge Dan (pseudonym), my main informant, who cleared a path for me at NewCar and served as an exemplar of social skill.

I am indebted to Steve Borgatti and Jason Davis whose collaboration on the 2014 *Research in the Sociology of Organizations* article laid the foundation for Chapter 1. They both also provided invaluable input on subsequent chapters and ideas as they evolved over several years. Similarly, many colleagues contributed valuable comments on chapters in this volume: Paul Adler, Phil Anderson, Chris Ansell, Paul Carlile, Francois Cooren, Charlice Hurst, Kate Kellogg, Paul Leonardi, Ajay Mehra, Martin Ruef, Stoyan Sgourev, Sameer Srivastiva, and James Taylor. Others offered critical input and support along the way: Wayne Baker, Evan Bernstein, Lorenzo Bizzi, Geof Bowker, J. P. Corneillson, Joe Ferrare, George Marcus, Gerado Okhuysen, Ray Reagans, Simon Rodan, Sim Sitkin, Mario Small, Beppe Soda, Leigh Star, Tom Steinberger, Sid Winter, Franz Wohlgezogen, Amy Wrzesniewski, and Pavel Zhelvazkov. Two anonymous reviewers provided extensive, crucial feedback. Arthur Goldhammer provided timely, insightful guidance regarding translations of de Tocqueville's *Democracy in America* in Chapter 7. My Cal State Fullerton colleagues, Shaun Pichler and Atul Teckchandani, reviewed and commented on countless drafts and passages—their input and support was invaluable.

When my commitment to the book wavered, I was buoyed by encouragement from Michael Cohen, who always took my ideas seriously from earliest days. On reading my proposal in late 2012, he wrote me: "The book plan is very ambitious. I find it very interesting [but] it will take a *lot* of writing effort. . . . To pull the variety of thinkers you have identified into what comes across as a coherent, intelligible argument will be a very big challenge. . . . If you can pull it off, it would be the kind of book that would . . . make [a] strong [contribution.]" His observations were a driving force in finishing the book and prescient regarding the challenges inherent in that effort. I hope this book achieves the impact he envisioned. His memory as a model scholar—brilliant, modest, interdisciplinary, and inclusive—is an ongoing inspiration.

Both Marc Ventresca and Darryl Stickel often served over time as vital sounding boards for messy early- and late-stage ideas, and countless iterations along the way, for which I am very grateful. I am especially grateful to JoAnn Horai, Ph.D., a dear friend and true scholar whose remarkable courage is a source of inspiration. Until her death in 2012, we often spoke daily about the most important ideas now found in this book. Her unblinking candor kept me on track. I profoundly regret that I cannot hand her a copy of the book.

As the manuscript neared completion, David Leibsohn dedicated a seemingly unending series of days (and nights) to ensure coherence, clarity, and consistency across numerous revisions of the chapters. I am also indebted to several people who came alongside the effort: Nazanin Tadjbakhsh for background research; Alex Toll, a rising junior scholar, for enormous help formatting the manuscript and valuable input on ideas; and Angela Belsky, who helped prep the final documents for submission.

At Stanford University Press, I am particularly indebted to Margo Beth Fleming for her long-standing encouragement, feedback, and guidance. She supported the book early and stood by the project for far longer than I had a right to expect.

I am grateful for the enduring love and support of my mother and father, both of whom inspired me on this journey. They each had a profound impact on my thinking and perception of the world. Sadly, both passed before the book's release, but it was a gift to be able to share with my mother the news that Stanford Press would publish the book, in the days just prior to her passing.

Finally, and most importantly, I want to acknowledge and express my heartfelt appreciation to those who shared in the everyday joys and burdens of this effort: Marcie, my wife, who provided support with patience and humor during the many hours away, while managing her own business and our family; and Rachel and Noah, my children, who tolerated my not infrequent absences and preoccupations while "working on the book." Thank you. I love you all very much and dedicate this book to you.

Getting New Things Done

Introduction

All larger organizations were once small. This seemingly self-evident statement raises the question, how do small organizations get bigger?

The answer I offer, briefly stated, is *by engaging in specific social processes through which they conceive of and pursue new outcomes.* Schumpeter described the creative response underpinning innovation and entrepreneurship as "getting new things done" (1947, 151). Getting new things done, in the form of distinctly novel change, is of special interest for its capacity to generate outsized impact in various contexts.[1] Of course, not all small new things lead to larger outcomes, but when they do, I will argue that it is frequently because skilled players mobilize both networks and knowledge to marshal support for markedly new initiatives.

In tracing organizational origins, we end up with variations on a pattern where a small number of strategic actors mobilize people and resources toward imagined future outcomes. Although every mobilization process is different, certain elements are common to the process of getting new things done across a wide variety of human endeavors, including business innovation, new venture growth, collective action, artistic movements, institutional emergence, and many other social domains. To be sure, there are often other external factors, such as natural disasters, institutional forces, legal or regulatory changes, or technological regimes, that may select, propel, and amplify these incipient efforts. However, in the end—that is, in the beginning—we need to explain how strategic actors drive growth along a predictable set of dimensions.

The Increasing Importance of Social Networks and Project-based Innovation

The ability to manage one's relationships, and the resources that those relationships afford, has always been a central concern of strategic actors. It's therefore worth examining how our networks and the means by which we orchestrate them have evolved over the past two decades. A series of digital tools for establishing, maintaining, and propagating ties, starting with mobile phones, e-mails, text messages, and more recently Facebook and LinkedIn, have dramatically altered the means by which we engage our networks. This was evident, for example, in the way that digital tools made the Arab Spring possible (Ghonim 2012). The importance of networking is also accelerated by the connectedness central to Friedman's (2005) account of globalization, which allows greater freedom and speed in the combination of people, ideas, and means for production.

Making the case that the way we engage networks has changed, Boltanski and Chiapello (2005, 2007) argue that society has evolved a "new spirit of capitalism," reflecting a shift from an industrial to a more project-based society, referred to as the "project-oriented city," involving a fundamental shift in organizing and individual action. According to the authors, the new way of organizing involves a firm "featuring an organisation that is very flexible; organised by projects; works in a network; features few hierarchical levels; where a logic of transversal flows has replaced a more hierarchical one" (2005, 165). The corresponding new form of individual action involves activity aimed at generating projects and formulating "life conceived as a series of projects" (169), which involves continually assembling disparate people for relatively short periods of time. This new form of individual action involves networks, coordinating, connecting, locating new sources of information, inspiring trust among those being coordinated, flexibility, and adaptability. Boltanski and Chiapello view key players in the new project-oriented context as "mediators [who] . . . possess the art of reconciling opposites, and know how to bring very different people together, and put them in contact" (2007, 115).

In the shift from an industrial world to a project-based world, new skills come to the fore. The ability for strategic actors to read the social terrain has always been crucial, but the importance of forming interdependent projects comprising multiple, interdependent actors has further amplified that talent. In addition, as people tend toward working less within the stable assemblages of corporations and more within constellations of projects, the ability to conceive of worthy projects, identify and assemble the appropriate participants and resources, forge trust in temporary communities, and cultivate ideas within professional communities constitutes a newly evolving toolkit necessary in a project-oriented world.

Coordination, Brokerage, and Social Skill

Mobilizing action to get new things done revolves around coordination, first and foremost of people but also of the resources or ideas they bear and the organizations they represent. In the absence of coordinative action, those elements stand largely inert. Coordination addresses how new things first get started through the novel combination and recombination of elements (e.g., as found in emergent start-ups and social movements), how new things get bigger (i.e., how elements accrue to a growing project or initiative), and how combinatorial elements come together repeatedly (through sustained feats of coordination found in organizational routines). My focus is on this microsocial crucible of action, usually involving a small number of strategic actors and some form of collaborative action among them.

Coordination involves the integration of interdependent tasks, and is therefore at the crux of organizing (March and Simon 1958; Faraj and Xiao 2006; Okhuysen and Bechky 2009). Coordination accomplishes this integration through the mechanisms of accountability, predictability, and common understanding (Okhuysen and Bechky 2009). Okhuysen and Bechky (2009) emphasize coordination within a design tradition that takes place in a single organization, in which participation, tasks, and outcomes are well established and relatively predictable. This formalized design tradition contrasts with more emergent contexts, where organizing is more creative, boundaries are less defined, and tasks and goals are unfolding. This underscores the need for an approach that is able to address not only predictable settings but also less predictable ones like the project-based contexts described above.

Coordination takes many forms. Weick (1969) introduced the double interact, a communicative exchange between two individuals, as the core coordinative unit of action crucial to organizing. In the double interact, according to Weick, one person communicates a message to a second person, to which the second person responds, followed by the first person making an adjustment to his or her original message based on that response. This three-step exchange within the dyad speaks to a broad range of communicative exchanges, such as when new ideas are pitched, existing ideas are refined, interests are gauged, or gossip is exchanged.

The double interact concept, though inherently coordinative, is characterized by a certain inertness in scale; it assumes a preexisting pair of interlocutors,
and considers primarily the possibilities for collaborative interchange within
the pair, but it does not address larger numbers of actors or the potential for
dynamic expansion beyond the two participating individuals. We could stretch
Weick's consideration of the dyadic interact to accommodate the exchange
between one person and a group (e.g., to which the person is presenting), or
between a leader and a larger audience. Nevertheless, the double interact still
fails to accommodate social phenomena inherent in feats of coordination, aggregation, and growth that involve a larger or growing number of participants
or more complex communication and response loops.

Imagine, as an alternative, that a member of that dyad introduced a third
person into the conversation. Such an initiation of a new tie introduces an important potential dynamism into the dyadic interact. That triad might then be
used as a proxy for a much wider range of dynamics associated with growth in
the number of parties engaged. When we bring a new person or organization
into a preexisting pair or examine an individual's introduction or facilitation
of two others, we invoke *brokering* activity—a central focus of this book. The
move from two to three, as Simmel pointed out over a century ago, is profound
in that it draws in a broad sweep of coordinative action of greater complexity,
impact, and dynamism (Simmel and Wolff 1950). The move to three speaks to
a number of social processes at the microsocial level, involving numbers far
greater than three (e.g., an emerging venture or an incipient episode of collective action), and, correspondingly, among firms of three or more at a more
macro level (e.g., alliance formation). In the simplest sense, the move from
two to three involves the alignment of a greater range of interests, ideas, and
resources.

To coordinate triads, a number of new social processes are implicated, including formation (e.g., the new introduction of nonacquainted others), in-
clusion (the invitation of a third into a preestablished group of two), and the
strategy or gamesmanship involved in growing groups (i.e., the strategic choice
to add or exclude a fourth or a fifth). This gamesmanship speaks to the fundamental dynamic of starting a social movement or organizing a dinner party. Because I am interested in the process of coordination that occurs among people
or organizations to get new things done, I will emphasize the social dynamics
in numbers of three and greater. This, I argue, is where an important dimension
of mobilization on a social level begins.

The more fully elaborated theory of brokerage that I present in this book offers a theoretical frame within which to understand the coordination in emergent collaborative action. Brokerage, as it is currently employed in the sociological literature, involves an intermediary (the broker) who stands between two others (or alters) who do not have a tie to each other. This "open" triad conceptualization of brokerage has a long tradition (Marsden 1982; Fernandez and Gould 1994; Burt 1992). In effect, we might characterize the classic definition of brokerage as involving two links and a gap. The links connect the broker to two parties, between whom the broker stands. The gap refers to the absence of a tie between those two alters. This brokerage formulation is important both for what it captures as well as for what it overlooks. The structural arrangement where the broker stands in between two alters is a simple yet powerful encapsulation of the relational issues central to the coordination puzzles noted above, and is of particular importance when actors seek to get new things done. This formulation is also important because of its second defining feature: the gap or disconnect between the two alters. As noted, a basic assumption in current research is that brokerage occurs exclusively in "open triads," causing scholars to overlook its equal relevance to closed networks, an assumption that unnecessarily obscures its connection to many forms of coordination and innovation.

This limitation unduly constricts many studies with a network focus. In an influential study of brokerage, Gould and Fernandez (1989) unpacked the brokerage phenomenon by identifying five empirically distinct "brokerage roles" distinguished by the different memberships (and associated interests) that the broker and each of the two unconnected alters might have. This approach allows for the evaluation of different structurally defined brokerage opportunities, but cannot empirically address the actual brokerage processes crucial to understanding coordinative phenomena (Spiro, Acton, and Butts 2013).[2] The problem with the structural approach, however, is that it is less articulate about the process crucial to understanding coordinative phenomena. I seek to explain how the brokerage triad, when combined with brokerage process, may be used to better understand how cooperation and coordinative action are induced.

Strauss points out that at its core, brokerage involves a representational act: a triadic process of "(1) a representing action, (2) with respect to some social unit, which is (3) directed at another unit (audience)" (1993, 172). In the move from two to three, the broker no longer engages with one other, as with Weick's double interact, but instead represents the position or viewpoint of one party (or World 1) to another individual or group (World 2). This, I will suggest, is far more than a special case of concern to social network theorists but the fundamental template for the coordinative act in organizing. A corporate manager advocating for an innovation within a firm, an entrepreneur mobilizing support for a new firm, or an activist initiating collective action on the street: each engages in different variations on these activities in triadic contexts involving numbers of three or greater.

The representational act described by Strauss inherent in triadic coordination is itself complex. Strauss observes that "the potential difficulty may be that the representing unit may not represent accurately (or honestly), or be judged not doing so by the represented" (1993, 172). That a strategic actor might intentionally represent inaccurately or adjust the knowledge ferried between alter parties has been anticipated by others (e.g., Burt 1992). There is, however, a raft of other communicative moves that the broker might make in selecting what might be extracted from one community and, once extracted, shaped for consumption by and communicated to another community. Strauss's (1993) observation that the broker's representation might not be judged as accurate suggests a range of cases whereby the broker continually shapes representations to maximize understanding, receptivity, or the perception of value—or to engender trust in the representation, the broker, or other combinatorial actors—as a prelude to collaborative effort. Taken together, these complex communicative moves suggest opportunities to leverage a structural orientation into a deeper understanding of the mobilization process.

Coordination between individuals or organizations becomes a more or less effective process in a social space due in part to a certain social dexterity. Such social dexterity might involve attributes like a command of the social hygiene specific to a given social context, the ability to forge trust between a broad range of familiar and less familiar actors, a talent for accurately assessing the character and interests of those whom you know less well, and a capacity to make effective appeals to enlist and connect people in the causes you wish to pursue. Examples of such dexterity can be found across recorded history. A combination of advantages arising from social network position and skill account, for example, for Cosimo de' Medici's ability to cultivate a position of power in fifteenth-century Florence (Padgett and Ansell 1993), Abraham Lincoln's ability to assemble support around legislation to abolish slavery (Goodwin 2005), or Paul Sach's creation of New York's Museum of Modern Art by assembling support from museums, universities, and finance (DiMaggio 1992).

Fligstein and McAdam define social skill as "the ability to induce cooperation by appealing to and helping to create shared meanings and collective identities" (2012, 46). They argue that social skill is an important tool for strategic actors, whether those actors are operating within existing orders or endeavoring to create new ones. Some actors, Fligstein and McAdam point out, are more skillful than others. Social skill, they further assert, requires a capacity to forge linkages within and across contexts. It also requires a general and a context-specific dimension, the latter of which may evolve if the rules and mores associated with various contexts change.

Based on their in-depth study of several historical cases, Fligstein and Mc-Adam (2012) identify several tactics that constitute social skill, including exercising direct authority, framing action, brokering (as noted, a core focus of this book), maintaining ambiguity, aggregating interests, and networking to outliers. While this list of tactics is certainly illuminating, it also suggests an opportunity to present a more parsimonious action framework consistent with the authors' emphasis on the strategic inducement of cooperation.

A Model for Emergent Collaborative Action

This book presents a microfoundational account of the network-based social origins of mobilization to present a theory of how new things get done. This microfoundational emphasis on small-numbers social phenomena has received increased attention in a number of disciplines including organizational theory and strategy (e.g., Barney and Felin 2013; Felin and Foss 2005), entrepreneurship (Ruef 2010), and collective action and institutional theory (e.g., Fligstein and McAdam 2012; McAdam, Tarrow, and Tilly 2001). The model I present draws on and contributes to many of the theoretical perspectives found in the organizational, sociological, and strategic literatures over the past years. The crucible of microsocial action I examine in subsequent chapters leverages work from a wide range of disciplines including social network theory, innovation studies, the knowledge-based view of the firm, entrepreneurship, pragmatism and symbolic interactionism, organizational routines, collective action, and sense-making. If in fact we have become, or always were, a project-oriented world, a well-elaborated theory of mobilization must draw on a broad set of inputs and correspondingly strive to present insight and implications associated with this crucible of action. In this book, I hope to link these differing literatures toward that effect.

The BKAP model—named for the model's three most distinctive features: brokerage, knowledge articulation, and projects—that I will explore in this book revolves around five key dimensions that can loosely be understood as two pairs of explanatory variables and one dependent variable, which is the mobilization of one or more focal strategic actors for some form of innovation I refer to as the *creative project*. The explanatory variables include *brokerage structure*, understood as varying primarily along an open or closed dimension (consider the open or closed triad discussed earlier); *brokerage process*, introduced initially from a network perspective that concerns the action by which a strategic actor leverages his or her network; along with the strategic actor's stock of *knowledge*, whether rooted in experience or education; and the strategic actor's *knowledge articulation* skill with which he or she communicates or articulates that knowledge for the purposes of engaging or enlisting others. The simple path model presented below (see Figure I.1) shows the basic relationships that define the BKAP model. I will briefly consider each of these variables below, and offer more in-depth treatment in subsequent chapters.

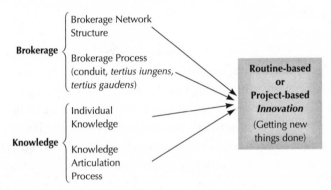

Figure I.1.   The BKAP Action Model: *Brokerage, Knowledge Articulation, Projects*
Source: Original figure.

The Dependent Variable: Innovation in the Form of Organizational Routines or Creative Projects

In most cases, networks and knowledge are harnessed for more predictable, incremental, or routine innovation. I argue, however, that the act associated with the mobilizing of uniquely new action involves a creative project, a microsocial unit of analysis I define "as an emergent trajectory of interdependent action initiated and orchestrated by multiple actors to introduce change into a social context" (Obstfeld 2012, 1571). In the latter case, the network and knowledge processes noted above may mobilize collective action in support of an innovation, a trajectory of multiple innovations, and in certain cases a newly emerging organization (or movement) catalyzed or defined by that innovation.

To present a clear understanding of the mobilization of action to get new things done, we must begin with some broader framework within which we can locate how actors engage action and, in particular, new action. All action, whether repetitive or oriented toward creating entirely new outcomes, exhibits some degree of choice or agency. Emirbayer and Mische (1998) argue that agency can be decomposed into three elements: the iterational, repetitive, or routine; the improvisational (or what they refer to as practical-evaluative); and the projective, involving the imagining of new alternative possibilities and acting on them. Taken together, these three forms of agency suggest that action may be conceptualized along a continuum characterized by the degree of repetitiveness and orientation toward the future.

Most of the time, actors are replicating or extending existing social orders and patterns of behavior. Stark information-processing constraints alone dictate that we can deliberate only on a relatively small percentage of our activity. As a result, routines and habit compose the vast majority of individual and organizational action. Even a revolutionary has routines, having to commute daily to the revolution, whether to the barricades in the city or the ramparts on the edge of town. How the revolutionary gets to work and how she goes about meeting with fellow revolutionaries bent on disruption is subject to some relatively mundane and repetitive activities. It is therefore not a surprise that one important theme in social movement theory concerns the importance of collective action routines and repertoires (e.g., Tilly 1976, 1993). As Strauss indicates, "Even in the most revolutionary of actions, the repertoire of routines does not vanish; at least part of it becomes utilized in combination with the new" (1993, 195).

A large number of innovation cases are also fairly routine. Consider the choice to roll out a new product in a product line or a new course at a university. In comparatively rarer moments, we break more substantively from routines to create an entirely new product, initiate a new protest, or from a more individual perspective, quit our job in order to start a new company or, more mundanely, switch from one coffee shop to another. Each of these actions illustrates a projective element that punctuates everyday routines. My emphasis is on those projective episodes that lead to sustained mobilization among multiple actors in the form of creative projects. To periodically mobilize new action in the form of creative projects, the BKAP action model argues that strategic actors engage the world through brokerage network structure, brokerage process, and knowledge along with knowledge articulation, the four explanatory variables which I turn to now.

Brokerage Structure

The first explanatory variable is an actor's social network structure, defined either by the pattern of ties that surround a given actor or that connect a network of actors together. Consistent with my brief treatment of brokerage above, networks can be either "open" or "closed." Open networks offer an actor an opportunity structure that is ideal for accessing new information but present the challenge of mobilizing disparate actors with diverse interests—a challenge I have characterized elsewhere as the "action problem" (Obstfeld 2005). Closed or dense networks, on the other hand, offer a focal actor a more interconnected network that is, on average, more homogeneous with respect to attributes and interests or more normatively constrained (Coleman 1988; Burt 1992; Obstfeld 2005). Dense networks therefore pose an opportunity structure where the surrounding network is more aligned and therefore more conducive to mobilization and coordination. That homogeneous network, however, is far less likely to be importing novel ideas and thus poses an "idea problem" associated with the greater likelihood of redundant information circulating within the network (Obstfeld 2005).

Of particular importance in tracing the distinction between closed and open networks is Burt's (1992) theory of structural holes. Consistent with the Marsden (1982) and Gould and Fernandez (1989) conceptualizations of brokerage as an open network, Burt defines a structural hole as "a separation between nonredundant contacts" (1992, 18). Burt offers an information and control argument for the structural hole advantage. Specifically, he suggests that brokers that stand between unconnected alters benefit both from the novel information that such a structure affords and from control benefits that allow the broker to leverage the disconnected actors. Over time, structural holes theory has become the predominant conceptualization of brokerage, because the structural holes-related measures have yielded compelling empirical evidence for the impact of structural holes on dependent variables at the individual and firm levels. The structural hole construct aligns with the properties of open networks. Burt equates brokerage with the open triad that defines the structural hole, and employs a contrasting term, "closure," to refer to the closed triad and the more closed networks with which it corresponds.

A network's properties are also determined by the content of the ties that connect actors, whether they involve friendship, advice, information exchange, or other relational content. Given a type of tie, the second determinant of a network is the structure of those ties as they relate to one another. Our treatment of brokerage thus far assumes an unnamed tie content and a structure: a broker who stands between two others where, as previously discussed, the alters may or may not have ties to one another. There are other tie and social network constructs, whether centrality (Freeman 1977, 1979) or structural equivalence (Lorrain and White 1971; Burt 1978), that we might consider, but I will restrict myself to network structures that directly implicate open or closed brokerage structures.

Brokerage Process

The second explanatory variable, brokerage process, concerns the way that brokers, whether individuals or firms, engage their network. I will focus on three variants of brokering: *conduit*, where the third party relays knowledge or information from one alter to the second alter without attempting to change the relationship between the alters; the *tertius iungens* (i.e., third who joins), defined as a strategic behavioral orientation toward connecting people in their social network, by either introducing disconnected individuals or facilitating new coordination between connected individuals (Obstfeld 2005); and the *tertius gaudens* (i.e., third who enjoys or benefits), where the broker exploits unfamiliarity, competition, or conflict between parties that the broker leverages actively or through purposeful inaction (Obstfeld, Borgatti, and Davis 2014). I will use this typology to develop a more detailed theory of how strategic actors mobilize support for initiatives.[3] The simplest rendering of this view is that the *tertius iungens* brokerage orientation presents the core mechanism by which strategic actors induce cooperation in order to get new things done. The more complex argument is that while *tertius iungens*-induced cooperation is the dominant theme in mobilizing action, social skill is a function of complex combinations of these three brokerage behaviors. We don't invite everyone we know to a dinner party but, rather, selectively invite those who, as a group, will make for the best social gathering (*iungens*), while choosing to exclude those we judge an inappropriate fit for the occasion (*gaudens*).

Knowledge

Knowledge and knowledge articulation constitute the second pair of explanatory variables. Consider that what any actor conceives as a goal, or a means for achieving it, is shaped by what he or she knows. Alongside the connecting of actors in a social network context, the organizational and entrepreneurship literature has recognized the importance of the combination and recombination of knowledge (Kogut and Zander 1992; Grant 1996; Spender 1996; Nahapiet and Ghoshal 1998; Hargadon 2003).

Individual stocks of knowledge are critical as a starting point for the recombining of knowledge central to innovation. Cohen and Levinthal (1990), for example, argue that a firm's capacity to recognize and assimilate information is a function of a firm's level of prior related knowledge. Their theorization about this capacity to locate and integrate new knowledge based on prior knowledge draws on research on knowledge acquisition at the individual level. In short, an individual's prior knowledge acquired through experience and education provides a critical resource, as well as a basis with which to evaluate and assimilate new knowledge critical to the pursuit of innovation.

Knowledge Articulation

The effectiveness of actors is also determined by their ability to communicate what they know to others whom they wish to enlist—a skill I refer to as knowledge articulation. One aspect of this communication skill involves the surfacing of tacit technical or social knowledge. Tacit knowledge refers to more unconscious, automatic, taken-for-granted understandings that are more difficult to surface or communicate (e.g., Collins 1985; Nonaka and Takeuchi 1995; Spender 1996; Gourlay 2004). In addition, in any social context where knowledge is in use, an actor must achieve intersubjective understanding of the matter at hand as a precondition to influence and enlistment. Polanyi (1958) used the term "articulation" to describe the communication of tacit knowledge but never defined it. Others have employed the articulation concept to reference the processes by which knowledge is made more explicit (Winter 1987; Benner 1994; Spinosa, Flores, and Dreyfus 1997; Zollo and Winter 2002; Dougherty 2004; Dougherty 2006). Building on this work, I define knowledge articulation as the social process by which knowledge is made more explicit, usable, or relevant to the situation at hand (Obstfeld 2001; Weick, Sutcliffe, and Obstfeld 2005). This articulation of knowledge is particularly important when people attempt to communicate some aspect of their tacit knowing for the first time, or move knowledge across a boundary in the pursuit of joint innovative action. In its various forms, articulation involves advocacy through talk (Mische and White 1998; Gibson 1999) and the active pursuit of intersubjective or shared understanding, which distinguishes it from a more general category of communication. This active orientation toward shared understanding is imperative, given the change in thinking, behavior, and even social order necessary to mobilize support around innovation.

Methodological Approach to Studying
How New Things Get Done

The elusiveness and idiosyncrasy of getting new things done makes the social scientist's task of locating and theorizing about it unusually difficult. To systematically study the social processes through which actors mobilize networks and knowledge, there are at least two empirical approaches. The first method, fieldwork or ethnography, involves direct observation and yields rich, in-depth insights about how new things begin. The value of fieldwork lies in its insight into the qualitative look and feel of previously misunderstood or overlooked phenomena. However, the risks and costs associated with direct observation are great, given the challenges of anticipating where to watch and how to watch with sufficient depth and rigor to come away with something useful to say. How do we recognize the sought-after newness when it occurs? What if an innovation is already underway when we begin our observation? If so, are we then still in a position to say something about how it got started? The central problem here is that the new doesn't officially "begin" at a single point in time. Rather, it is always preceded by some other antecedent conditions and events. There is rarely an Adam-and-Eve moment in the progression of technological or social change.

A second approach is to select an already-large organization or movement of interest and work backward historically (e.g., McAdam 1990; Tilly 1978). Such an approach presents the scholar with the obvious advantage of selecting in advance the larger organization or movement outcome that the study of incipient conditions will predictably yield. The historical approach, however, poses other methodological predicaments, such as the danger of sampling only on successes or failing to capture the true origins or mechanisms of social phenomena begun some time ago and likely far away. The historical approach may yield synoptic clarity, but often at the expense of unpacking the precise social mechanisms central to the specific mobilization process in question. There is also important insight to be gained from combining insights derived from closer-in field observation and survey work with examples from historical analysis, and from sampling briefer cases of different kinds of beginnings. Such a blended approach yields a broadened perspective on emergence, an illumination of the twists and turns that lead (or fail to lead) to bigger things, salient founding figures with strong intentions and visions of future states, and perhaps most importantly, specific mechanisms and practices that make for getting new things done.

This book presents a parsimonious, broadly applicable theoretical framework drawn from these multiple sources to account for how new things get done. To portray how networks and knowledge are mobilized, I will draw on my own field observations, quantitative survey research (sometimes in the same setting), and other research and examples from a broad range of social phenomena and academic literatures. While I began this project with extensive ethnography of a single organization, I subsequently employed supplementary methods and analysis to derive a theory with much broader application to the range of organizing.

Field Observations Illustrating How New Things Get Done

My essential argument is that the microsocial mobilization of shared action has a set of dynamic, interlocking characteristics common to organizational innovation, entrepreneurship, collective action, and transformation of institutional fields, among others. To illustrate such a dynamic social process, I begin with a brief vignette drawn from my extensive field observation and survey study of an automotive design process (Obstfeld 2005, 2012) involving the efforts of several strategic actors, primarily middle managers, collaborating to reengineer the prototype parts-purchasing (PPP) process at a major U.S. automotive manufacturer, which I refer to as AllCar. The initial instigators of the mobilization effort were Brian, an AllCar project manager, and Dan, the program manager for the build of an entirely new vehicle, the G5.[4] Brian and Dan were both long-term employees socialized into a homegrown AllCar "activist" tradition that arose from the organization's several decades' effort to innovate and bootstrap in the face of substantial resource constraints. The scrappy, risk-taking heirs to this self-described "cowboy" culture formed a loose network who shared an understanding and experience of how to surreptitiously leverage their network to assemble a critical mass of support before going above ground with a more formal advocacy for a given innovation. Dan once said of Brian,

> Brian is a mole. He's a gopher. He's an underground player. . . . He's partnering. He's out there [burrowing] around. . . . Brian is underground with Johnson [a high-ranking manager in engineering operations]. He's underground with Carl. He's underground with Brad of [the] GreatCar [division].

Dan, whom I had observed for hundreds of hours, displayed the same underground collaborative behaviors. This loose cluster of cowboys had achieved notable success. In one case, for example, they were instrumental in pushing the organization to develop and employ the world's first three-dimensional automotive design process for a production automobile.

When I first spoke with Brian, he was preparing to orchestrate a cross-division initiative to redesign the company's antiquated, inefficient PPP process. Dan, in league with Brian, had also roughed out a somewhat synchronized scheme to leverage his senior management position and network in response to similar issues identified by the AllCar group that his G5 division was also encountering. Dan's plan was to redesign the G5's prototyping procedure, while simultaneously creating and staffing a new G5 PPP unit, under his direction, to operate it.

The two efforts met very different fates. Brian's more ambitious AllCar group, after an underground planning phase, was initially successful in mobilizing broader and higher levels of support. Dan's narrower, within-division plans, while encountering periodic resistance, ultimately succeeded as he assembled several constellations of lateral and upper-level support while skirting procedural roadblocks and resistant stakeholders. Critical to Dan's (and the AllCar group's) efforts was his ability to fashion a clear vision and logic through which he recruited several high-ranking executives, as well as the actual team members for the new PPP unit, before it was approved. Dan's compelling "pitch" for forming a new G5 PPP team, which he refined in numerous meetings and backroom conversations, led to a repeated reference to the importance of "getting the right parts to the right car at the right time." The success of AllCar's and Dan's efforts hinged on intimate knowledge of the AllCar community and its subgroups, the capacity to mobilize networks across corporate departments and professional ranks—a mobilization that was made possible by the protagonists' ability to translate their ideas into compelling, well-timed appeals that often successfully enlisted support, shaped that support into collaborative efforts, and defused opposition. "Exogenous" factors also had a continual role, ultimately derailing the AllCar effort and periodically disrupting and altering, but not sinking, Dan's initiative. Though Dan's implementation constituted a local success (whose subsequent merit was yet to be tested), the same network dynamics were responsible for several previous AllCar innovations, some of which resulted in broad impact across the organization and in some cases the industry.

I use this illustration to bring to life, in brief form, the nature of mobilization, which I will elaborate in far greater empirical and theoretical detail throughout this book, but also to point out the family resemblance of this innovation-focused effort to mobilization efforts in other domains that also involve local action motivated by a desired future state pursued by actors skilled in enlisting and connecting others in projects of growing scope and impact. For example, similar dynamics are found in most early-stage entrepreneurial efforts. As an alternative to a research tradition that has tended to focus either on new-firm creation as an event or on heroic individual entrepreneurs, Ruef recently proposed an "emerging organization" perspective that begins "with the intuition that startup efforts . . . involve collective action that is oriented toward the founding of a new organization." According to Ruef, "entrepreneurs, in this conception, are defined by their intention to form a social group" (2010, 7). This microsocial view of entrepreneurship is captured in Ruef's (2010) illustration of group formation, in which an entrepreneur, Luis Hernandez, persuades two partners to start a wholesale clothing business. Much like Dan at AllCar, Luis's initial formation of this ownership core, along with his subsequent enlistment of support, constitutes the same initial nexus of collaboration found in many start-ups. In this case, the necessary but frequently overlooked feature of start-ups, even to some extent by Ruef, is the connecting activity that entrepreneurs like Hernandez orchestrate between entrepreneurs, investors, and other sources of support (e.g., employees and advisors) that provides resources necessary to the firm's growth.

In the social movement literature, McAdam, Tarrow, and Tilly argue that brokerage is one of the core mechanisms in collective action, where brokerage involves "the linking of two or more unconnected social sites by a unit that mediates their relations with each other and/or another site" (2001, 142). That common social processes are invoked here is no accident, though what remains to be offered is a more precise microsocial account of these social processes with application to multiple domains. Fligstein and McAdam (2012) offer a compatible grasp of the microsocial origins of institutional fields. It is such an account that I hope to present here.

Structure of This Book

The structure of this book is as follows. In Chapter 1, I introduce a more detailed account of brokerage process, which I describe as composed of three basic brokerage orientations toward action. I then explain how strategic actors employ these different action orientations, to provide a more detailed account of how mobilization is accomplished. In Chapter 2, I explore knowledge and knowledge articulation as complementary resources and skills. Collaborative action takes place within the field of social networks, but the discursive practices associated with knowledge articulation are a critical means by which coordination is accomplished. I describe the interaction between networks and knowledge articulation with ethnographic data. In Chapter 3, I provide a more in-depth explanation of the mobilization outcome—creative projects—and its contrast to organizational routines. While "getting new things done" is evocative, a more rigorous account of nonroutine creative outcomes is necessary. In Chapters 4 and 5, I provide in-depth ethnographic data to illustrate how strategic actors interweave brokerage and knowledge articulation in the pursuit of routine-based innovation (Chapter 4) and nonroutine innovation (Chapter 5) in an automotive engineering and design environment.

Having framed this basic model for the mobilization of action, I take a deeper look in Chapter 6 at the theory of using social skill to mobilize innovative action, drawing on and extending Mead's symbolic interactionist perspective and examining insights from fieldwork regarding perspective taking and use of multiple voices (riffing). In Chapter 7, I explore how the BKAP model can be employed to explain creative outcomes in artistic movements, entrepreneurship, collective action, and several other issues in organization theory and strategy. I have also alluded here to unique challenges posed by digital tools and contexts. It is now not unusual for a layperson to assume that any reference to social networks primarily alludes to Facebook and LinkedIn. I explore how digital contexts shape and reshape mobilization processes and brokerage using the Arab Spring as an illustration. On a more practical note, I then reflect on opportunities that social skill as conceptualized here may have for addressing pressing issues associated with education, social inequality, and social mobility. I conclude by arguing, in the spirit of de Tocqueville, for the "science of association" as the master social science.

My emphasis on process and structure is reminiscent of the stance taken by Weick (1969) almost fifty years ago in his book *The Social Psychology of Organizing*. Weick's book title involved a small alteration to the title of another well-established book, *The Social Psychology of Organizations*, by the famous social psychologists Katz and Kahn, also at the University of Michigan. Weick's understated but clever title change suggested a shift from viewing organizations as structures to organizing as a process. In a similar approach, I attempt to consolidate a shift already implicit in some social network literature, from social networks as a structure to the brokerage processes that constitute network behavior. The objective, of course, is not to abandon the structural view, but to build a complementary, and ultimately more complete, theoretical perspective. In the past, a focus on "social networking" was often suspect as a province of hackneyed truisms and idiosyncratic ruminations developed in the absence of data or theory. Nevertheless, I argue that this process-based perspective presents valuable new insights for understanding both mobilization as it has unfolded in historical contexts, and the accelerated collaborative environments that characterize the world we live in today.

1

Brokerage in Action

Determining how actors get new things done requires us to consider both *social network structure* and *social process*. As described in the Introduction, this kind of collaborative action grows out of *brokerage*, the process by which actors are strategically assembled and coordinated, and by which—in some cases—participation is expanded.

Concepts of social network structure provide the architecture and accounting of such activity, while concepts regarding social process explain the formation and aggregation of ties among people or groups (or the absence thereof). Both are of fundamental importance to social network theorists. Yet in social network research, the study of social process is often subordinated to structure, because structure is easier to measure and, in many cases, process is inferred from network data. Such a stance diverts attention from theorizing about network process. It is that imbalance that I seek to address. Building on an emerging stream of research, I will identify various network processes that are important in dynamic organizational phenomena, such as technological innovation and entrepreneurship (Obstfeld 2005; Vissa 2012; Long Lingo and O'Mahony 2010; Davis and Eisenhardt 2011; Bizzi and Langley 2012).

Given its central place in organizing, brokerage requires a sound theoretical foundation. The traditional social network perspective, as noted in the Introduction, defines brokerage in terms of open networks, open triads, or structural holes (Burt 1992). The open triad represents the most "extreme" challenge that

might confront the broker: coordinating two parties who were heretofore unconnected and who may have come from unconnected social worlds. But as I
will discuss, the coordinative challenge found in the open triad also occurs in
closed triads, just as the ongoing challenge to establish intersubjective meaning
is found in all dyadic communication (Weick 1979; Parsons 1951; Luhmann
1995).[1] Consequently, brokerage phenomena are found in all groups of three
or greater—whether in open or closed networks—and in the more complex
networks that combine open and closed characteristics that are most often
found in the real world.

   This chapter addresses social network structure and process to explain how
brokerage functions to get new things done. I present this perspective in a sequence of five steps. First, I frame innovative action as often unfolding in triads
through brokerage. I am aware that this may strike many network scholars as
obvious, but I believe that it is an "obviousness" worth reviewing, as the triad
is so often and so easily overlooked. Second, I explain how network structure
sets the context for action, by summarizing some of the well-established ways
in which network theorists analyze social network structure, emphasizing the
distinction between open and closed social network structures. Third, I clearly
distinguish between brokerage as action and brokerage as structure. Fourth, I
consider in more depth the brokerage process, first by defining it and then by
proposing three fundamental brokerage orientations or behaviors: conduit, *tertius gaudens*, and *tertius iungens*. Finally, I revisit Fligstein's idea of social skill, or
the ability to induce cooperation (Fligstein and McAdam 2012; Fligstein 1997,
2001), to argue that inducing cooperation to get new things done is achieved
by strategically combining the three brokerage orientations toward action.

Innovative Action Unfolds in Invisible Triads

For the network theorist, the role of the broker may seem obvious, and yet the
richness, ubiquity, and importance of triadic or coordinative phenomena may
be easily overlooked. The coordination of the broker, whether momentary or
ongoing, is often forgotten or overlooked altogether by the parties being connected, by observers outside of the coordinative unit of action, and even by the
broker himself. We often note the formal, routinized role of brokers in facilitating transactions such as the sale of stock or the transfer of real estate, but we
overlook the deceptively complex coordinative work associated with bringing
parties together. If we focus our attention on the informal, "in-between" broker roles in which people continually engage, the number of cases to consider

increases dramatically. A parent serves as a go-between in the tense relationship between her spouse and her teenage son.[2] A biology professor brokers between her students and the scientific community. A head of state endeavors to resolve a point of tension between two other heads of state.

The term "invisible work" (Star and Strauss 1999; M. C. Suchman 1995) has been used to denote how certain kinds of coordinative, uncounted, or uncountable work may be ignored, devalued, or dismissed altogether.[3] Conversely, work that is more closely associated with established forms of expertise and status is more likely to be registered, counted, and valued. There is a rough analogue here to the attention that social network analysis pays to structure, which is easier to measure as compared to network process. Triadic coordination, too, is more difficult to observe because it is more spread out over time and space as compared to more localized individual or dyadic phenomena. More simply put, triadic social processes are inherently more complex and, therefore, more difficult for either the casual observer or the social scientist to detect and track.

The broker's formal and informal coordinative work—consequential yet hard to detect—extends to firms, organizations, or larger social entities such as nations. For example, the venture capital firm links an entrepreneurial organization to investors and other sources of talent the new firm may need to engage. Similarly, brokerage relationships connect the supply chain, in which each firm in the chain brokers between its suppliers and downstream buyers by transforming inputs into outputs, and thereby adding value. In this sense, all commerce is inherently triadic, from Wal-Mart to the hot dog vendor. Mergers and acquisitions may appear to be dyadic pairings of the acquiring and acquired, but in fact they do not proceed without the intercession of brokers before the deal (to locate the participating parties and negotiate terms), during the deal (to plan and facilitate the strategic and operational integration), and after it (to knit together information technology departments and reconcile organizational charts and cultures). The network theorist emphasizes the ubiquity and importance of these triadic patterns of coordinative action, in contrast, for example, to Weick's dyadic interacts.

As the merger and acquisition example above suggests, brokerage also involves mixed cases of individual and organizational actors. In the example, we find that individuals constitute a critical unit of analysis that stands between two or more merging organizations to facilitate the knitting together of the senior officer, operations, information technology, and marketing clusters as well as two organizational cultures and strategic visions. Individuals, either employed by the merging firms or consultants, reconcile, bridge, and knit together various

aspects of the merging organizations. In a similar spirit, a diplomat brokers between individuals and organizations, governmental or otherwise, from the host
country, on the one hand, and a corresponding set of entities from his or her
country of origin, on the other, in countless variations on a basic triadic scheme.

Each of these coordinative feats involves different degrees of effort, attention, local knowledge, experience, and skill. Although all brokerage action involves agency exerted over time, there are distinctions to be made between the
kind of brokerage that occurs largely within a routine versus that which occurs
within a creative project (Obstfeld 2012), a distinction I will explore in more
depth in Chapter 3. Connecting an interested stock buyer and seller is nearly
automatic as there exists an active ongoing market mechanism for supporting such transactions, whereas facilitating a productive meeting between two
embattled spouses or unfriendly heads of state is far more difficult. Expectations between the parties to the stock sale are commensurable, and sale procedures are highly routinized. The meeting between the heads of state, on the
other hand, might confront not only their deeply rooted antipathy and distrust
but, quite likely, opposition to their dialogue from many stakeholders that they
represent.

In many cases, the brokerage triad (i.e., the broker and two alters) is less visible than an individual's reputation, talent, credentials, or the more easily grasped
dynamics of dyadic exchange. The (seemingly) dyadic exchange of a job interview obscures the preceding introduction that made the interview possible. We
move to the foreground the most immediately apparent dyadic pairing, but we
do so at the expense of the brokerage activity that made that pairing possible.

Network Structure Sets the Context for Action

A key objective of social network analysis is to get at how larger patterns of
ties enable and constrain action (Gulati and Srivastava, 2014). The first independent variable in the BKAP model is the broad category of social network
structure—in other words, the formal "plumbing" or pipes (Podolny 2001)
that define relationships among social actors. The basic building block of social
networks is the relationship, or tie, between actors. From this fundamental unit,
patterns of ties can be identified that reveal opportunities for and constraints
upon action.

Ties vary by their strength (i.e., strong, weak, or somewhere in between)
and their content (e.g., friendship or work collaboration). One has close friends

and work colleagues and, less frequently, close friends at work, and all manner of variations on these parameters, including relationships that feature other tie content, such as advice, mentoring, and information exchange. Ties often involve multiple forms of content, or multiplex ties, as in the case of a tie involving both friendship and work collaboration. Various network measures use tie data to evaluate different facets of network structure. Network degree centrality, for example, is a relatively simple measure of the number of ties that a given actor has—an actor with more ties has greater centrality. Betweenness centrality (Freeman 1977) measures the number of times an actor stands on the shortest path between two other actors in the network.

Open versus Closed Network Structures

The open versus closed network is an analytic dimension of particular relevance to how brokers access knowledge and mobilize action to get new things done. The well-established debate around the relative merits of open and closed networks (e.g., Marsden 1982; Burt 1992, 2004; Krackhardt 1999; Ahuja 2000) bears directly on considerations of network structure and process. Open networks, featuring the absence of connections among those in the network, as noted earlier, afford greater access to novel, nonredundant information and greater discretion to act (i.e., less normative control). Earlier, seminal network research associated with the flow of novel information concerned related but not equivalent network structures involving weak instead of strong ties (Granovetter 1973). Closed or dense networks allow for more efficient communication, greater trust, social support, and normative control (Uzzi 1997; Hansen 1999; Coleman 1988). Coleman (1988) illustrates the virtues of such closed networks with the example of a mother who moves from Detroit to Jerusalem because, among other reasons, her unattended children at play will be looked after by adults in the vicinity: an advantage of a cohesive network at a community level. Similarly, Coleman (1990) illustrates the mutual trust characteristic of certain dense networks by citing the willingness of tightly knit London diamond dealers to exchange diamonds for examination without contracts or documentation.[4]

These two social network types pose different opportunities and challenges for two key aspects of combinatorial innovation: new ideas and the coordinated action to implement those ideas. Open networks present both a *knowledge advantage* and an *action problem*. By situating people at the confluence of different social domains, open networks create opportunities for accessing new ideas and combining them in new ways. These same networks, however, may create challenges for *acting* on such ideas, because the dispersed, unconnected people across different social domains are inherently more difficult to mobilize or co-ordinate, especially around novel ideas.

Closed or dense networks, conversely, reduce the obstacles to initiating co-ordinated action necessary to implement innovation (an *action advantage*) but pose greater obstacles to the generation of new ideas (a *knowledge problem*). Networks that are closed or dense are conducive to mobilized action, because interests and perspectives are prealigned or normatively constrained, and the language and trust necessary to mobilize those interests are more readily available (Granovetter 2005). Dense networks, while presenting optimal conditions for the efficient exchange of the complex but shared knowledge necessary for innovation in complex organizations (Uzzi 1997; Hansen 1999), present a knowledge problem, because of the redundancy of information circulating within the network (Granovetter 1973).

A variation on the simple open-closed distinction involves what I will refer to as hybrid cases. To understand the hybrid case, consider first the strongest example of the cohesive, dense network at a triadic level, consisting of what David Krackhardt (1999) refers to as Simmelian triads: those in which each of the three actors involved have strong reciprocal ties to one another. Now, by contrast, consider variations on the closed triad, in which two of the actors exhibit weaker ties with each other, perhaps reflecting newly formed relationships. Finally, consider a larger network with a focal actor (often referred to in the social network literature as "ego") who has many more alters than the two that would constitute a triad, and blends weak, strong, and absent ties of diverse content. These blended networks serve as the basis for most network research from which we must extract insight regarding how various social processes drive meaningful outcomes.

An important line of social network research, for instance, explores how a combination of network cohesion (i.e., a measure similar to density) and range (i.e., ties to different knowledge pools) facilitates knowledge transfer and diffusion, and thus represents an important form of social network advantage (Reagans and Zuckerman 2001; Burt 2002; Gargiulo and Rus 2002; Reagans, Zuckerman, and McEvily 2004; Reagans and McEvily 2008; Srivastava, 2015). Fleming, Mingo, and Chen (2007), while finding evidence for the advantage of open networks for supporting the novel combinations that define creativity, also present evidence that closed networks accelerate the sharing of data and feedback, along with better distributed understanding and ownership of ideas. As we press the open-closed distinction to its limits, we eventually bump up against the need to speak to the action that takes place within hybrid networks that display both open and closed properties.

The limits to the open versus closed social network distinction are suggested by what Aral and Van Alstyne (2011) refer to as the diversity-bandwidth tradeoff. While open networks may provide access to more diverse information and serve as a source of more novel information, Aral and Van Alstyne point out that the weak ties often found in open networks may only support the exchange of "simple news." In contrast, they find that the stronger, higher-bandwidth ties found more frequently in dense networks and associated with greater mutual understanding, more frequent contact, and higher trust, may actually foster greater exchange of richer, more complex, and potentially more sensitive information, and consequently greater learning. This is consistent with Coleman's (1990) argument that strong and reciprocal ties with others increase access to information. Similarly, Ter Wal et al. (2016) found that open networks combined with shared professional specialties that provided similar knowledge and experience helped interpret diverse information, and alternatively, the limitations of closed networks can be augmented by shared third parties who can correct for overly narrow interpretations (see also Reagans and McEvily 2003; Rodan and Galunic 2004).

These recent empirical studies point out the limitations of simple distinctions between open and closed networks. These distinctions risk overly simplified inferences that open triads automatically serve as better sources of new information and closed networks function as echo chambers of redundant information. It is for this reason that sophisticated analysis of patterns of ties demands a concomitant grasp of the social dynamics associated with those ties. The importance of brokerage process is suggested by Burt, who asserts, "Broker advantage does not result from getting diverse information so much as it results from personal abilities exercised and developed while engaged in diverse information" (2013, 14).

The limitations to the open-closed social network distinction were initially surfaced by my social network study of innovation in the automotive design environment introduced in the Introduction. In that context, I observed engineers pursuing innovation by actively coordinating work across boundaries, often by skillfully connecting interests across multiple departmental boundaries.

Initially, I assumed that my extensive daily observation of automotive engineers' design activity took place in open networks characterized by an absence of ties among the various alters the skilled actors engaged (i.e., many structural holes) and that such activity, in turn, led to more creativity and innovation on the automotive platform. I was therefore surprised to learn from my subsequent social network analysis in the same automotive division that the combination of dense or closed networks along with what I refer to as a *tertius iungens* orientation, an action-based measure of actors' orientation to connect others in their networks, predicted innovation involvement (Obstfeld 2005). To be sure, the importance of dense networks in this context was shaped by the automotive technological landscape I was observing, involving incremental innovation with large clusters of collaborating engineers tasked with designing error-free designs and safe, high-volume, production vehicles. In this technological context, *tertius iungens* connecting activity and closed networks were both predictive of innovation, but in other contexts it could easily be open networks and *tertius iungens* connecting activity that account for innovation. Nevertheless, the odd juxtaposition of closed networks and action focused on connecting others forced me to rethink my assumptions about the automatic link between network structure and behavior and therefore to disaggregate structure and process.

Distinguishing Brokerage as Action from Brokerage as Structure

A wealth of organizational research ratifies the concept of brokerage as the network structure associated with the open triad or structural holes (e.g., Barley 1996; Pollock, Porac, and Wade 2004; Fleming, Mingo, and Chen 2007; Xiao and Tsui 2007). The open triad conceptualization of brokerage has a long tradition. Marsden suggests that brokerage is a mechanism "by which intermediary actors facilitate transactions between other actors lacking access or trust in one another" (1982, 202). Similarly, Fernandez and Gould indicate that brokerage is a "relation in which one actor mediates the flow of resources or information between two other actors who are not directly linked" and underscore that brokerage "does not permit the endpoints of the brokerage relation to be directly connected" (1994, 1457). Consistent with this tradition, Burt defines a structural hole as "a separation between nonredundant contacts" (1992, 18) and has characterized the open-closed distinction as involving "brokerage and closure" (Burt 2005). His empirical work (e.g., 1997, 2004) makes a compelling case for the importance of the open triadic social network structure.

Burt (1992) offers a theoretical underpinning for the advantages that accrue to actors with many structural holes in their networks and for the mechanisms by which those advantages are secured.[5] He suggests that brokers who stand between unconnected alters benefit both from the novel information that such a structure affords and from control benefits that allow the broker to leverage the disconnected actors against one another. This latter argument, in turn, draws on the work by Simmel regarding the *tertius gaudens*, or "third who enjoys" (Simmel and Wolff 1950). A broker with a *tertius gaudens* orientation enjoys benefits passively by not intervening in the conflict or disconnection between two alters, or more actively by playing alters off one another. Over time, structural holes theory has become the predominant conceptualization of brokerage, because various structural holes-related measures have generated empirical evidence for the impact of structural holes on dependent variables at the individual and firm level, and because the persuasive triadic theoretical argument applies to networks at the individual and firm level (Burt 1992).

The limitations of the reliance on the open triad to define brokerage can be illustrated with two examples. In the first case, imagine Jack, a professor in a business school. Jack walks down the hall to invite Sally and Jane to join him for coffee, intending to discuss a potential collaboration. Sally and Jane know each other well and are part of an accounting department whose faculty have collaborated in every conceivable combination, to the extent that all faculty members have coauthored papers, in one form or another, with every other faculty member. A similar relationship might be seen in a product development group or a loosely formed community of musicians.

I argue that Jack's effort should be viewed as brokerage, despite the absence of a structural hole between Sally and Jane. Thus, working within a closed or dense network with an absence of structural holes, in which one individual undertakes coordinative action that generates some new collaboration, still constitutes brokerage. Going a step further, perhaps in initiating collaboration between Sally and Jane, Jack deliberately leaves out Joe, who is also connected to Sally and Jane. Jack may have decided not to include Joe because of a past collaboration that did not turn out well, or because he fears that Joe's talent might put Jack's leadership of the new initiative in jeopardy. Here again, we have a case of brokerage behavior (i.e., *tertius gaudens*) taking place within a nominally dense network. By relaxing the central criterion for brokerage currently in use—absence of ties between alters—we generate a new set of cases in which coordinative action by a third (or *tertius*) can be fruitfully considered.

(See, for example, Kaplan, Milde, and Cowan's [forthcoming] study of boundary spanning within closed networks in a university interdisciplinary research center.)

In a second case, consider an open triad in which the broker, Jack, has ties to two unconnected actors, Matt and Deborah, but never does anything that involves linking across or leveraging the disconnection between the two ties, and within which, more generally, no social process has occurred. If no such coordinative action occurs, is it useful to call Jack a broker? The identification of structural holes within a network does not necessarily imply any specific social activity. The presence of structural holes may create a pronounced opportunity for brokerage, but as suggested by the first case, brokerage can occur without structural holes. The potential for brokerage lies in the broker having ties to two or more parties, not in the ties or lack of ties among those parties.

I argue for expanding the theoretical terrain to make a distinction between strictly structural patterns (such as structural holes) that Burt and others have associated with brokerage, and the social behavior of brokering. There are two parts to this argument. The first is the recognition that brokerage can occur in a wide variety of structural contexts, including closed or dense networks. The second is a separation of motivation and opportunity, a distinction that Burt deliberately downplays: "I will treat motivation and opportunity as one and the same . . . a network rich in entrepreneurial opportunity surrounds a player motivated to be entrepreneurial. At the other extreme, a player innocent of entrepreneurial motive lives in a network devoid of entrepreneurial opportunity" (1992, 36). It is often the case, however, that even when a given structural pattern affords opportunity for brokerage, the intent and effort of brokering will vary. Social network structure affects the ways that brokers do their brokering, but does not define it. Consistent with Obstfeld, Borgatti, and Davis (2014), I propose a reworking of the Marsden definition of brokerage that accommodates the social process that occurs within a social network context but that may unfold somewhat independently of the network itself.

After Obstfeld et al. (2014), I revisit Marsden's definition of brokerage as a mechanism "by which intermediary actors facilitate transactions between other actors lacking access or trust in one another" (1982, 202) as a point of departure for an expanded definition. The expression "lacking access or trust" corresponds to the non-tie condition in Burt's definition of structural holes but is broader, in that it acknowledges that the nodes being brokered may have some kind of tie with each other (just not a sufficient one). I propose to adapt Marsden's definition to accommodate brokerage process in both open and closed networks by modifying three of the above terms: "transactions," "intermediary," and "lacking access."

First, instead of "transaction," I will use "interaction" to capture a broader range of social activity beyond the cross-sectional or economic exchange that the word "transaction" suggests. Such interactions address brokerage activity that includes the broker who moves knowledge from one alter to another (Gould and Fernandez 1989), the broker who leverages conflict (Simmel and Wolff 1950), and the broker who introduces or facilitates the engagement of two alters (Simmel and Wolff 1950; Obstfeld 2005). Interaction accommodates not only the kinds of discrete events in time that are suggested by transaction, but also a relational pattern of engagement over time, in which a broker may facilitate tie formation or the growth of trust or friendship between two other parties.

To disaggregate the brokerage definition from a particular structural pattern, I refer to the broker as an "actor" as opposed to an "intermediary" (Marsden 1982), because "intermediary" implies that the two parties with whom the broker engages do not have a tie to each other, certainly an important category of brokerage activity, but not the only kind. Stated differently, I do not associate Marsden's "lacking access" condition to the absence of a tie but, rather, suggest that it can also occur in the presence of an alter-alter tie. In light of these considerations, after Obstfeld, Borgatti, and Davis (2014) I simplify and broaden the Marsden definition of brokerage to the following: *behavior by which an actor influences, manages, or facilitates interactions between other actors*. Note that this definition retains a structural element in that it limits attention to situations with three or more parties. Note also that the revised definition adds "influencing" and "managing" to Marsden's original "facilitating." This denotes a broader range of activity that different forms of brokerage might involve, and opens the door for a more complex consideration of brokerage process.

Three Brokerage Orientations Toward Action

Thus far in this chapter, I have treated different forms of brokerage interchangeably, focusing on the triadic nature that they share. Now, after Obstfeld et al. (2014) and Obstfeld (2005), I will differentiate among three basic categories of brokerage process (Figure 1.1): conduit, *tertius gaudens*, and *tertius iungens*. These distinctions, introduced briefly in the Introduction, will be considered more fully in the pages that follow. Whereas conduit is more concerned with knowledge transfer, *tertius gaudens* involves more competitive behavior, and *tertius iungens* concerns more collaborative or facilitating behavior.

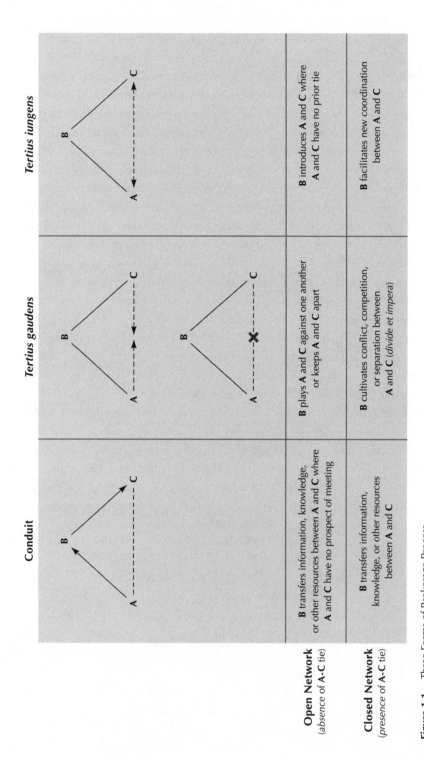

Figure 1.1. Three Forms of Brokerage Process

Source: Obstfeld, Borgatti, and Davis 2014, 142.

Conduit Brokerage

At its root, conduit brokerage is a knowledge transfer phenomenon involving the passing of information between parties (Burt 2004; Obstfeld 2005; Obstfeld et al. 2014). In conduit brokerage, as the term suggests, the third party relays information from one alter to the second alter without attempting to change the relationship between the alters. The diffusion of innovations is often a conduit process. As Everett Rogers (2003, xx), the diffusion of innovation scholar, indicates, "The diffusion of innovations is essentially a social process in which . . . information about a new idea is communicated from person to person." Simply put, the broker wishes to move a chunk of knowledge or information (e.g., an idea, an innovation, a story, a practice), which I will refer to as the "token," from Alter C to Alter A.[6] The source or destination can be an actual person or a more general context. Given this, moving forward I may replace "Alter C" with "World C" or "Organization C," on the basis of the specific context of a given example. I indicate alter or world to suggest that the conduit knowledge source or destination can be an actual person or a more general context.

The simplicity of the knowledge transfer framing, however, belies the potential wrinkles and complexities involved with moving the token. The framing neglects characteristics related to the broker, the token, and the alters. Neglecting these characteristics fails to fully appreciate the key dimensions necessary to successfully move a token from one alter to another. One has to identify, discover, or create the token "found" in World C, move the token successfully to World A (and potentially to other worlds), and possibly charge for its delivery. The token can range from a discovery of substantial new knowledge or technology to an incremental insight that might enhance a preexisting project.

One conduit issue concerns the broker's level of intentionality. The conduit broker may stumble upon a new token in World C that triggers the realization of an opportunity associated with bringing the token back to World A where it is unknown. According to one account, Howard Schultz stumbled into both the idea of coffee culture and the caffè latte drink when visiting Milan, and brought both back to Starbucks (Giannoni 2013). Alternatively, the broker may engage World C with World A (or a project in World A) in mind, that is, in search of ideas he can bring back to World A. A corporate executive may attend a conference with the objective of finding a solution to a particular problem in mind; a marketing executive may attend an event with an eye out for ways to reinvigorate her company's stale product line. This suggests a level of social or entrepreneurial alertness (Kirzner 1989; Hayek 1948) characteristic of some brokers. The extent to which such ideas are discovered or created

(Alvarez and Barney 2005) constitutes a major debate within the entrepreneurship literature.

As part of conduit brokerage, a broker curates from among ideas and opportunities to determine which will be of interest and should therefore be passed on to one's target audience, and *how* that transfer should occur. Whether involving the transport of major discoveries or smaller embellishments, conduit brokerage necessarily reformats knowledge and information for successful transit and reception. A broker who learns of a new technological breakthrough in one social domain must be poised to deliver that new knowledge to a second actor or domain unfamiliar with that knowledge (Hargadon 2002). Within this curation activity, there is sensemaking that involves the noticing, bracketing, and labeling of discovered ideas or opportunities (Weick, Sutcliffe, and Obstfeld 2005) in social space and time (Strike and Rerup 2016).

Conduit-based knowledge movement involves more sustained agency and translation than is often associated with knowledge transfer in the organization, strategy, or social network literatures. In the 1970s, during his visit to Xerox's innovation lab, Xerox PARC, Steve Jobs was shown the graphical user interface and mouse. According to one observer, Jobs perceived Xerox's GUI and mouse technology as far more than a feature enhancement: "He was very excited. . . . when he began seeing the things I could do onscreen, he . . . started jumping around the room, shouting, 'Why aren't you doing anything with this? This is the greatest thing. This is revolutionary!'" Subsequently, he ferried these tokens back to his own company, Apple Computer (Gladwell 2011, 44). Consider how Jobs's movement of mouse knowledge from Xerox back to Apple had elements of search, discovery, embellishment, reconfiguration, and as some have asserted, theft. (This also suggests how even theft may have a seat at the conduit brokerage table.)

The visit by Jobs and his Apple engineers to PARC represents the first leg of the brokerage triad, with the second leg involving his return to Apple with the idea. That second leg involved a reconfiguration of the discovered opportunity. According to the engineer with whom Jobs spoke upon his return to Apple, Jobs's instructions were as follows: "[The Xerox mouse] is a mouse that cost three hundred dollars to build and it breaks within two weeks. Here's your design spec: Our mouse needs to be manufacturable for less than fifteen bucks. It needs to not fail for a couple of years, and I want to be able to use it on Formica and my blue jeans" (Gladwell 2011, 46).

The way that the knowledge or information which the conduit broker encounters in World C and subsequently ports back to World A can also be understood in terms of the aspiration levels described by Cyert and March (1963). Once organizational performance falls below the aspiration level, firms are more likely to engage in a problemistic search for solutions and organizational changes (Argote and Greve 2007). As the conduit broker encounters different information in World C, she makes judgments about the relevance of this information for issues found in World A to which she will return—judgments that might influence her aspirations for existing projects or routines and for the potential initiation of new ones, along with consideration of the transfer and translational moves necessary to successfully transport that knowledge back to A.

Finally, the broker's activity involves the potential motivation to extract a reward or rent (Burt 1992; Kogut 2000). The broker may demand, or hope for, a reward or rent in exchange for supplying information. The motivation of the broker can also manifest in choosing to move information but with different choices about whether and how much to charge. A broker might, for example, ferry multiple novel social facts from one community to the other without any immediate or eventual fee. This latter case suggests that the rents potentially associated with conduit brokerage are not automatic and rely on the motivation of the broker (e.g., to help, to increase his or her status, or to make money) as well as the ability of the broker to extract rents.

The broker's intent may be to extract short-term profit or develop long-term market access (Pollock et al. 2004; Hallen 2008); pursue profit at an individual or collective level (Kacperczyk, Davis, and Hahl 2011; Ryall and Sorenson 2007); and may range from strict self-interest to more complex combinations of individual, shared, or communal objectives (Klein et al. 2006; O'Mahony and Ferraro 2007). Conduit brokerage intent might range from altruistic and rent-free knowledge facilitation to exploitation (e.g., by theft of information) or extortion. Moreover, the broker may forgo charging rent due to a calculative rationality that forgoes some form of rent for unspecified gains in the future, a consideration I will return to with respect to the *tertius gaudens* and *tertius iungens*.

The broker's movement of knowledge justifies closer scrutiny. Carlile (2004) addresses the complexities of knowledge movement with a model that suggests that sharing knowledge across boundaries involves knowledge transfer, translation, or transformation, depending on the extent of the differences between the boundaries (aka alters) being crossed and the newness of the knowledge being moved.[7] For Carlile (2004), transfer involves the simplest knowledge-movement challenge of crossing a syntactic boundary, in which the differences between actors are known and there is a common lexicon that is sufficient to move the token across the boundary. Translation is necessary when the broker must move knowledge across semantic boundaries where different interpretations exist because the contexts in which people develop their knowledge are different (I will return to transformation later). Carlile's work provides a pointed critique of the traditional knowledge transfer literature's oversimplified model of knowledge transmission. This critique of the information-processing perspective underlying knowledge transfer also targets the literature's underestimation of the effort required in many cases of knowledge movement, a critique consistent with the often unrecognized invisible work found in triadic action noted earlier.

Jobs's transport of the GUI-mouse token back to Apple involved both transfer (e.g., describing the functionality he observed) and translation (e.g., "here is how this technology would make sense in the Lisa and Macintosh computers, and this is what we need to do next to make it work"), in order to move knowledge across the organizational boundary between Xerox PARC and Apple. In the basic story, however, getting new things done required a conduit-like engagement, consisting of identification of potential opportunities across a boundary (between Apple and Xerox PARC), followed by the transfer and translation necessary to move the knowledge associated with the opportunity back across the boundary to Apple. In other cases, the knowledge "discovered" in one world might be ferried to one of many other worlds (other than "home") that the broker encounters.

Conduit brokerage, and the idea of moving the token between two worlds, is most consistent with the knowledge advantage associated with open triads. But structural holes are not a prerequisite, as we can readily imagine situations in which alters are already tied in some way and yet conduit brokerage is a common occurrence. Let us return to the hypothetical business school described above, in which Jack is a professor who has ties with Sally, who does research on organizational alliances, and Jane, who studies innovation. In a conversation with Sally about her recent paper submission, Jack learns of several of Sally's papers that are relevant to Jane's study of innovation and which he subsequently brings to Jane's attention. The likelihood that Jack will bring novel information from Sally to Jane when all three have ties is certainly less than in the case where Sally and Jane have no ties, but that outcome is still possible and consequential. This suggests a specific case in which conduit brokerage may take place in a dense network of ties. More generally, the movement of knowledge within dense networks remains an area of intense research interest (e.g.,

Lee, Bachrach, and Lewis 2014), underscoring the importance of considering conduit as a brokerage behavior occurring in both closed and open networks.

Conduit brokerage is a frequently studied brokerage form. This is not surprising, given its correspondence with established themes associated with knowledge transfer in the literatures of strategy, organization theory, and the diffusion of innovation. Less attention has been paid to the related brokerage processes *tertius gaudens* and *tertius iungens*, even though they are equally important to a process-based view of brokerage. I turn to them now.

Tertius gaudens Brokerage

Tertius gaudens refers to situations in which a broker maintains or exploits unfamiliarity, competition, or conflict between parties actively or through purposeful inaction. *Gaudens* behavior can range from the exclusion or separation of parties to the active fomenting of conflict. This brokerage orientation, first articulated by Simmel (Simmel and Wolff 1950) and later explored by Burt (1992) in connection with structural holes theory, involves a strategic intent and effort to generate advantage presented by the disconnection between two parties.[8] Speaking to the motivation of the *tertius gaudens*, Burt notes, "The *tertius* plays conflicting demands and preferences against one another and builds value from their disunion. . . . When you take the opportunity to be the *tertius*, you are an entrepreneur in the literal sense of the word—a person who generates profit from being between others" (1992, 34).

I summarize three basic expressions of *tertius gaudens*, two of which are explicitly noted by Simmel. The first form of *gaudens* noted by Simmel involves the broker playing alters off one another. Podolny and Baron provide an example: "When two individuals are suppliers or buyers of the same resource, a third individual can exploit the competitive relation between the other two to play them off against one another. Even if two individuals are not vying for or proffering from the same resource, the third can exploit the lack of connection by inducing a competition for his or her time" (1997, 674–75). As Burt (1992) points out, the *gaudens* strategy may involve rivals in pursuit of the same relationship, as in the case of two or more buyers who want to buy the same object, or simultaneous demands made by alters in separate relationships with the *gaudens*. Alter-alter antagonism may be strong, and the potential for contact between alters also poses the risk that they could align to eliminate the *tertius gaudens*'s leverage, and even conspire together against the *tertius*. As Simmel indicates, "The favorable position of the *tertius* disappears quite generally the moment the two others become a unit—the moment, that is, the group in question changes from a combination of three elements back into that of two" (Simmel and Wolff 1950, 160). A variation on this *gaudens* strategy would involve playing hypothetical or fictitious alters against one another. As illustrated again in Podolny and Baron (1997), the broker may claim that he or she has other unnamed suppliers that will provide a lower price.

A second, more subtle but still substantive form of *tertius gaudens* involves the broker's choice to keep alters apart who have no immediate opportunity to meet or coordinate. In such a case, the *gaudens* leverages or preserves some form of unfamiliarity between alters. This second iteration might come into play when the broker declines to invite someone to a critical meeting or social gathering, due to competitiveness or considerations of compatibility or timing. There is an implicit connection between the conduit and the *tertius gaudens* activity. While distance, language barriers, and a variety of other inefficiencies and incompatibilities often render the conduit broker's introduction of the source of information to the recipient infeasible (i.e., one simply cannot introduce every information source to everyone the broker informs), the broker often actively avoids such introductions to preserve the novelty and value of the information she ports from one alter to the other.

I include as a *gaudens* form Simmel's consideration of an explicit cultivation of conflict, which he refers to as *divide et impera* (divide and conquer). The *gaudens* and *divide et impera* brokerage orientations resemble each other, in that they both involve the broker's exploitation of alter-alter tension. In *divide et impera*, the broker actively encourages conflict between alters: "The distinguishing nuance consists in the fact that the third element intentionally produces the conflict in order to gain a dominating position" (Simmel and Wolff 1950, 162). For parsimony, I consider *divide et impera* as a variant of the *tertius gaudens* orientation, as it involves neither conduit behavior nor the joining behavior associated with the *tertius iungens*, which we will examine shortly.

The role of information movement, central to conduit brokerage, is often also key to the *tertius gaudens* (and *tertius iungens*, explored below). In her movement of knowledge, the broker may not always strive for the clearest or most accurate message but may instead resort to some form of obfuscation. Burt's observation that "accurate, ambiguous, or distorted information is strategically moved between contacts by the *tertius*" can evolve into information strategies where information is altered or withheld to keep alters apart or encourage conflict (2000, 355). In all brokerage behavior, the broker manages the flow of information, though it is more likely in the *gaudens* case that it will involve blocking or distorting information, with the broker as a disproportionate beneficiary. In this sense, while the goal of the conduit is to traverse the boundary (and for the *iungens* to reduce or eliminate the boundary), the goal of the *gaudens* is to reinforce the boundary or, at the very least, to leave it in force.

While the disconnection between alters leveraged by the *tertius gaudens* often assumes the absence of an alter-alter tie (Burt 1992), there are numerous exceptions to such an assumption. In the competing buyers example noted above, the buyers might be quite aware of and even know each other and have extensive familiarity with each other's competing offers and demands. In the case of Simmel's *divide et impera*, the active conflict fomented by the broker constitutes a negative tie between the alters and suggests the existence of a prior tie of some kind. Burt suggests the tension between the presence and absence of ties in the *gaudens* case: "Successful application of the *tertius* strategies involves bringing together players who are willing to negotiate, have sufficiently comparable resources to view one another's preference as valid, but won't negotiate with one another directly to the exclusion of the *tertius*" (1992, 33). Such a specification of *gaudens* behavior suggests the possibility of some form of relationship between alters. Similar tension is found in Burt's use of Merton's role-set concept to suggest a *tertius gaudens* behavior where the player at the center of a negotiation assigns to competing members of the same role-set the task of resolving their contradictory demands (Burt 1992, 31). In such an example, alters clearly have ties to one another (and as will become clear in the discussion below, this activity even suggests a *tertius iungens* linking behavior).

To summarize, the competitive posture found in Simmel's *tertius gaudens* behavior appears to have validity independent of a structural condition and may involve cases where competing alters are aware of or actually have a relationship with one another. The *gaudens* orientation does, however, rely on the disconnection of alters and, in this sense, aligns with the alter-alter disconnection most salient in the open triad. In a roughly similar sense, conduit behavior occurs frequently in both open and closed networks, although the clearest illustration of both the conduit's and *gaudens*'s properties, advantages, and translational challenges are found in the open triad.

There is an immediacy and concreteness to the "profit" to be generated by the *tertius* in pursuing a *gaudens* strategy that seeks to create an immediate tension between alters, with the impact on a specific outcome (e.g., a price, a decision) at stake. In other cases, there may be a longer-term objective to create separation between certain players and various sources of information or opportunity. This can be contrasted with the longer-term strategy of building trust and reciprocity by facilitating information or alter-alter ties more associated with the *tertius iungens*, considered below.

Tertius iungens Brokerage

The last basic form of brokerage considered here, *tertius iungens*, involves the broker's introduction or facilitation of two other parties. Where the *gaudens* leverages disconnection or competition, the *iungens* actively pursues coordination and connection between alters. In previous work (Obstfeld 2005), I have suggested a distinction between brief *iungens* and sustained *iungens*. Brief *iungens* refers to interactions involving discrete episodes of introduction, in which the broker introduces or facilitates ties between parties, and a continuing coordinative role is unnecessary, diminishes in importance, or is simply not offered. Sustained *iungens* is where the broker's ongoing facilitation is required.

While the *tertius iungens* takes Simmel's treatment of the nonpartisan as a precedent (Obstfeld 2005), Simmel's nonpartisan is concerned only with the reconciliation of tensions between antagonistic parties and does not explicitly consider the case of the *tertius iungens* introducing previously unconnected alters. As Simmel puts it,

> The non-partisan either produces the concord of two colliding parties, whereby he withdraws after making the effort of creating direct contact between the unconnected or quarreling elements; or he functions as an arbiter who balances, as it were, their contradictory claims against one another and eliminates what is incompatible in them. (Simmel and Wolff 1950, 146–47)

Network expansion, whether in the form of entrepreneurial start-ups or emerging social movements, is likely to rely on this connecting of previously unconnected parties. Due to the *tertius iungens*'s role of leveraging open triads or disconnection by creating connection and closure, the *iungens*'s behavior is uniquely and somewhat paradoxically positioned between and actively engaged with both open and closed networks. In previous work (Obstfeld 2005), I illustrate how the *iungens*'s introduction of unconnected alters eliminates one structural hole (creating closure) while creating downstream structural holes (open networks) in ego's second-tier network (Figure 1.2).

Imagine a case where Fred endeavors to initiate collaboration around a creative project (Obstfeld 2012) by enlisting Gloria and Joann. In the simplest case, Fred presents the project idea at a lunch meeting with Gloria and Joann in the form of a story that describes the project's origins, objectives,

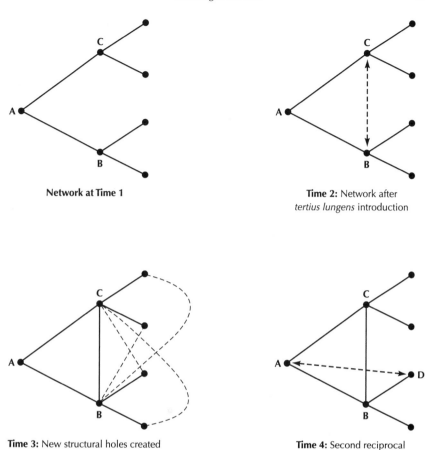

Network at Time 1

Time 2: Network after
tertius Iungens introduction

Time 3: New structural holes created

Time 4: Second reciprocal
tertius Iungens introduction

Figure 1.2.   How *tertius iungens* Activity Creates Structural Holes
Source: Obstfeld 2005, 122.

collaborative dynamics, and successful conclusion, with the associated enhance-
ments to wealth and status. Alternatively, Fred may sense that a one-step, in-
person meeting is premature given the lack of complete alignment between
Gloria's and Joann's interests. As a result, Fred may alternatively approach Gloria
first with a version of the proposed project, story P_G, tailored to appeal to Gloria's particular interests and concerns, and then approach Joann with a second
version of the project story, P_J, tailored to Joann. With these two successful
dyadic interactions in hand, and the potential for increased trust with both parties, he now introduces the project at the lunch meeting with Gloria and Joann
through a third version of the project, P_{GJ}. Story P_{GJ} reflects what Fred learned

in the previous two exchanges and is tailored to maximize the joint appeal to Gloria and Joann. Not surprisingly, Fred may choose to begin with the points of greatest shared appeal to both Gloria and Joann and omit altogether those issues particularly objectionable to either. Fred's communicative activity certainly involves translation but also, quite possibly, transformation and creation of new knowledge (and opportunities) that transcends each party's local interests.

In such a scenario, we can imagine additional wrinkles that involve the continual enhancement of the project story as appeals to new alters are made or new story features, with broad appeal to the expanded set of alters, are discovered. Consider the following act of diplomacy between Barack Obama and the heads of state of France (Sarkozy) and China (Hu Jintao) (Cooper 2009). According to the *New York Times* (April 2, 2011), Sarkozy and Hu Jintao, in a large conference room surrounded by eighteen other world leaders, began sparring over tax havens. The news report indicates that Obama first approached one leader, then the other, before bringing the two together to resolve the dispute:

> According to accounts provided by White House officials and corroborated by European and other officials also in the room, Mr. Obama escorted both men, one at a time, to a corner of the room, to judge the dispute. How about replacing the word "recognize," Mr. Obama suggested, with the word "note?" . . . It was not a Middle East peace accord. But Mr. Obama had his first moment as a statesman.

As the examples here suggest, *tertius iungens* brokerage might occur in cases where alters already know one another and in other cases where alters are unconnected and brought together for the first time.

Where conduit brokerage motivation might range from exploitation to altruistic and rent-free knowledge facilitation, there has been a tendency to see the *gaudens* as self-interested and the *iungens*'s act of joining as more altruistic. The altruistic associations with *iungens* are salient when associated with the behaviors by which transformational leaders may facilitate the coordination, efficacy, and professional advancement of their staff through various forms of network facilitation (Grant 2013), or on the international level when *iungens* connecting work facilitates public health cooperation at an international boundary, thus countering the divisiveness exerted on transnational networks by state institutions (Collins–Dugral 2012). That said, it's important to remember that *tertius iungens* connecting is also a means by which strategic actors get new things done. The *tertius iungens* may forgo charging rent for coordinative services due to a calculative rationality that anticipates unspecified gains in the future. People for whom longer-term collaboration appears to be a greater probability, that is, where there is a reasonable chance that actors will meet again and therefore the shadow of the future looms larger (Axelrod 1984), may have a greater propensity to see the potential benefits of playing the *tertius iungens* and thus feel more motivated toward connecting others.

The earlier case of Jack, Sally, and Jane illustrates how *tertius iungens* behavior occurs within both open and closed triads. As with the conduit, the features of the *tertius iungens* work are most salient when introducing previously unconnected actors (i.e., the open triad), but this connecting work is equally relevant to coordinative work in denser networks. References to *tertius iungens* brokerage often assume an open triad or structural hole as a necessary condition; however, the potential for *tertius iungens* brokerage to occur in either dense or sparse networks is anticipated in my definition: "a strategic behavioral orientation toward connecting people in their social network by either introducing disconnected individuals or facilitating new coordination between connected individuals" (Obstfeld 2005, 100). In the definition's first case, the *tertius* introduces disconnected individuals: a structural holes case. In the definition's second case, however, the *tertius* facilitates coordination between previously tied individuals. The brief and sustained *iungens* examples, taken together, suggest the tension between the presence and absence of ties. In the case of the brief *iungens*, the introduction of alters suggests the elimination of the alter-alter structural hole. The sustained *iungens* case, however, suggests that despite a connection between alters, some aspect of alter-alter disconnection may endure that necessitates sustained engagement of the *tertius* to secure the *iungens* brokerage interaction.

Per the definition above, the *tertius iungens* construct is referred to as a strategic behavioral orientation to denote an action preference for approaching problems in a social context (Higgins 1998; Levine, Higgins, and Choi 2000). I view the *iungens, gaudens,* and conduit as brokerage orientations toward action, but given their strategic nature I will alternatively refer to them as strategic orientations or strategic orientations toward action. Note that "orientation" does not imply an internal motivational or psychological state but an orientation toward acting in the world.

Orchestrating Use of Multiple Brokerage Orientations

While brokerage in some cases involves a choice from among one of the three brokerage orientations (e.g., Shi, Markoczy, and Dess 2009), it often entails a combination of conduit, *iungens*, and *gaudens* behaviors. Broker repertoires consist of different combinations of these behaviors. Multiple brokerage behaviors might be pursued simultaneously in different parts of the broker's network or might evolve from one to another over time. As noted earlier, skilled brokerage often involves selective deployment of these approaches with different actors, under different contexts, or for different objectives. Every dinner party, for example, involves invitations extended to some (which in aggregate constitute *tertius iungens* brokerage) and the choice to exclude others (*tertius gaudens*) deemed inappropriate for the occasion. The reasons such invitations are extended or withheld may range from compatibility of the invitees to the number of seats at the dinner table.

Tertius iungens behavior is often enabled by prior conduit and *gaudens* episodes. So where the major thrust of the brokerage may involve *iungens*, conduit and *gaudens* often facilitate the *iungens* act. Consider the choice of timing: I might not introduce you to my friend Kareem (one leg of the triad) until I have a better feel for your talents and trustworthiness (the second leg of the open triad). The non-tie between you and Kareem is the product of a *gaudens* orientation I may consciously maintain. At some time in the future, however, arranging an introduction might make sense in connection to a developing project I am pursuing; in other words, a shift over time from *gaudens* to *iungens*. Alternatively, Quintane and Carnabuci (2016) use longitudinal e-mail data to show how broker networks evolve from conduit to *iungens*.

Extended Illustrations

The strategic brokerage orientations can sequence in the opposite direction. Long Lingo and O'Mahony's (2010) study of independent country music displayed combinations of *gaudens* and *iungens* brokerage: *iungens* first, to generate creative possibilities for songs, followed by *gaudens* to curtail search and focus on song production. Of course, a multistep sequence is possible: in another case, Long Lingo and O'Mahony illustrate a *iungens–gaudens–iungens* sequence involving a producer, Hank, who tries to connect three record labels with his still under-the-radar artist. He plays the three labels off one another and then introduces one of the labels to the artist:

> I got the showcase moved to a larger club. I got the head of the [Century] record label and the head of A&R and another label [Fantasy Records] to the showcase. **[producer as *tertius iungens*]** And they [the band] were smart

enough to bring a truckload of fans to plant in the crowd. So when they hit the stage the place went unglued. And the [Century] record label people are going, "What the hell is this?" And I'm just chuckling to myself, "This is amazing." The head of A&R, he said to me, "Did they write that song?" And about two songs later, "Did they write that one too?" Then he looked back at [Fantasy label head who was also present]. Inviting [multiple] label heads to the same concert created competition **[producer as *tertius gaudens*]**.[9] (62)

Ultimately, the producer's creation of a *gaudens*-based competitive market for this new artist's talent leads to a brokered deal between the performer and the label:

So then I went back and talked with them [Fantasy] a bit. I could see they were digging it. . . . They said, "Who else knows about this?" "Well," I said "[Century] is here." **[producer as *tertius gaudens*]** He [Fantasy Records] said, "I want to sign them tonight." **[producer as *tertius iungens*]** (62)

This vignette also suggests how one broker's (the producer's) *tertius iungens* strategy can co-occur with or lead to another broker's (the band's) *tertius iungens* strategy (i.e., mobilizing a truckload of fans to plant in the crowd). Additionally, the convening of a highly attended venue through a *tertius iungens* strategy also sets the stage for the *gaudens* strategy that follows. Notice, too, how the brokerage involves an interplay between individuals and, in some cases, the organizations that they represent.

Venkatesh (2013) illustrates the interplay of brokerage strategies in a very different context involving gang activity in a Chicago neighborhood, while also demonstrating ancillary brokerage behaviors. These include "hiring of a *iungens* broker," sequences of *tertius iungens* behavior (which I refer to as *tertius chains*), and interesting alter stipulations necessary for them to participate in a *tertius* engagement (which I refer to as "alter preconditions"). Venkatesh describes the following scenario:

In this neighborhood . . . there's a young man named Johnny and he shoots William, shoots him over an altercation, they're both teenagers. So what happens after this shooting? Johnny's family decides that they need some help. They're nervous because Johnny and William don't know each other. In fact, the parents of Johnny don't know the parents of William, they don't know William, so they contact a school teacher and they say, "Can you help figure out, since these two people are probably in your school, how we can get them together or solve this problem? We have a shooting." **[enlisting teacher as *tertius iungens* broker]** The school teacher says, "Well, I can only do some part of it, but I might need some help," and so she calls this Pastor Wilson, who says, "I probably could help you as well, let's bring the two groups together." **[*tertius* chain; *tertius iungens* 1 enlisting *tertius iungens* 2]** Johnny says he won't come until his street gang leader, whose name is Tito, allows him to enter a meeting of mediation **[alter precondition]**. Tito says, "I have to come along." OK, so he has to come along. So William says, "Well, my leader has to come along." **[alter precondition]** So a gangster disciple gang that William is a part of, that leader comes to the meeting as well. So now it's [the] pastor, it's the teacher, it's Johnny, it's William, and these two street gang leaders talking about what happened in the shooting. (Venkatesh 2013)

Another of Venkatesh's ethnographic cases illustrates multiple brokerage strategies at work, including a novel prosocial form of *tertius gaudens*:

Jeremiah shoots Tino and their [respective] families are concerned. . . . Their family calls the police. . . . **[enlisting broker whose role is unspecified]** The police have to engage a formal process of finding out, conducting an investigation, etc., but he [a policeman] also decides that he is going to call a clergy member and that clergy member is going to come in and start meeting with the two young people. **[broker enlisting *tertius iungens*; *tertius* chain]** The clergy member says, after meeting with the two young people, finding out how hostile they are, that he calls the four gun traders that he knows in the neighborhood— the gun brokers—who sell guns in the informal market and he tells them, "I don't want you selling anymore to either Jeremiah or Tino for the next 30 days because I need to talk to them; I need to quiet them down." **[broker enlisting *tertius gaudens*; *iungens/gaudens* broker]** But there are more gun brokers in the neighborhood. So he calls somebody who used to work with the Chicago Transportation Authority who's retired who knows the other gun brokers in the neighborhood, and that person calls two other folks, and says, "OK, guys, don't sell to Jeremiah and Tino." This is resolution. It looks like madness perhaps, but this is an attempt at mediation in this neighborhood. **[*tertius* outreach to facilitate *tertius gaudens*; *tertius* chain]** (Venkatesh 2013)

I argue that what Venkatesh refers to at the end as "mediation" masks what the brokerage typology reveals to be a set of brokerage action sequences, which can be analyzed for greater insight in terms of the multiple-skilled and less-skilled paths by which brokerage processes can unfold.

My use of "tertius chains" above refers to how a sequence of *tertius* opportunities is set up by *tertius iungens* introductions. When a broker makes an introduction between alters, she creates a new potentiality in the network. One *tertius* introduction transforms each alter's network and affords new opportunities for action that would not otherwise be available. In referring to this as *tertius* chains, I adapt Harrison White's (1970) concept of vacancy chains. In one prominent example of White's vacancy chains (1970), an individual leaves his current position to take advantage of the new vacancy, which in turn creates a new vacancy, which is then filled by another actor. A vacancy chain ensues, filled by actors who subsequently create new vacancies until the sequence terminates. Whereas vacancy chains are based on cascades of vacancies, *tertius* chains are based on cascades of *tertius iungens* introductions. This reference to *tertius* chains foreshadows the construct of creative projects explored in Chapter 4.

Sequenced patterns of brokerage activity can be found in every form of organizing, and much of the work on multiple brokerage behaviors has been conducted in corporate contexts. Davis's (2011) study of innovative alliances in the computer industry found that active pruning of old ties (*tertius gaudens*) may be necessary before managers can effectively facilitate new ties, suggesting that sequences of *gaudens* and *iungens* behavior are sometimes necessary. Ozcan and Eisenhardt (2009) found that becoming a broker in the mobile gaming industry requires simultaneous *iungens* activity, where two alters are coupled with the threat to disconnect either party in order to motivate both parties to be brokered, suggesting that the threat of *gaudens* can induce *iungens* in real time. In an ethnographic study of the implementation of health care reform in two community health centers, Kellogg (2014) found that brokers engaged in *tertius iungens* brokering around high-status, high-value reform work that was well aligned with their professions, but distanced themselves by engaging a buffering *gaudens* brokering in low-status, low-value reform less aligned with their professional training.

Most recently, Burt, Merluzzi, and Burrows (2013; Burt and Merluzzi 2016) found that sequences of engaging and disengaging from closed networks, which they refer to as "serial closure" or "network oscillation," provide greater advantage than similarly structured networks that are consistent over time. Such episodes of closure might emerge, according to Burt et al. (2013), from reversals that result in diminished status. These closure episodes might alternatively stem from the temporary pursuit of creative projects (Obstfeld 2012) that bring together clusters of similarly focused actors. Taken together, these empirical cases demonstrate how effective brokerage strategies may require complex combinations and sequences of different brokerage behaviors and how skilled actors may command repertoires comprising multiple brokerage behaviors for this purpose.

Social Skill

The brokerage strategies described above comprise the three basic approaches to coordination. "Social skill" (Fligstein 2001) refers to the differential abilities of strategic actors to induce cooperation, mobilize action, and otherwise direct shared action toward a social outcome. As noted in Chapter 1, Fligstein argues that such social skill is unequally distributed. He further observes that to make the idea of social skill empirically useful, one must specify the behaviors that socially skilled actors use to induce cooperation. Fligstein (1997, 2001) identifies a set of somewhat idiosyncratic tactics: direct authority, agenda setting, taking what the system gives, framing action, wheeling and annealing, maintaining "goallessness," aggregating interests, networking to outliers, and brokering (understood here as mediating between competing parties). He also references Mead's symbolic interactionist work, stressing how coordinated action is accomplished through the forging of shared meaning.

I argue that the strategic use of the three brokerage behaviors is the underlying action framework for Fligstein's tactical repertoire and constitutes a more fundamental locus of social skill. To fully explain social skill, however, we must also explore the single, sequential, simultaneous, and parallel deployment of the three different brokerage strategies in use.

The Advantage of *tertius iungens* for Getting New Things Done

Consistent with Obstfeld et al. (2014), a central intuition here is that within complex and dynamic social settings consistent with a more project-based world, brokerage process, as opposed to brokerage structure, is of increasing importance to brokerage. In such settings, structural advantages are more difficult to maintain and leverage, while deployment of brokerage behaviors provides a means for adapting to these multifaceted and rapidly evolving circumstances (Goldberg et al. 2016).

A second intuition is that while skilled social actors often simultaneously, sequentially, and in parallel deploy the three brokerage orientations, which are found in all acts of coordination, *iungens* behavior becomes increasingly important in responding to the demands of newly emerging intra- and interorganizational work that is flexible, networked, and organized in projects (Obstfeld 2005; Boltanski and Chiapello 2007; Hagel, Brown, and Davison 2010). As I have suggested, the *tertius iungens* activity offers the opportunity to create new combinations as an alternative to the control of the *tertius gaudens*:

The activity of the *tertius iungens* is most challenging when the nature and prospects for projects are uncertain, and the relevant actors to engage are not apparent. In these cases, identifying the different actors to engage and the appeals that will resonate with those actors are the subject of considerable skill, quite discrete from the structure of the social network itself. (Obstfeld 2005, 122)

Conduit brokerage similarly provides value to one group by porting needed complementary resources from another. It is more restricted because it merely allows for transfer of knowledge or resources without the opportunity to integrate and leverage complementary resources in developing a more novel solution, precisely what is required for getting new things done. *Tertius gaudens* brokerage largely relies on the similarity of alters to be effective. Simmel suggests that *tertius gaudens* is suited to conditions where two vying parties "keep one another in approximate balance" (Simmel and Wolff 1950, 157). The substitutability of the alters from the broker's perspective, in terms of both type of tie and type of alter, is what allows the two alters to be played off of one another. Ultimately, a *gaudens* orientation is more likely to avoid or block interactions and combinations.

What do I mean by "complex and dynamic social settings"? Consider a given network as a cast of characters with four basic properties: (1) identity (who the cast members are); (2) the size of the cast; (3) the relationship between the various cast members (the raw data from which various network properties are determined); and (4) the resources, be they knowledge or wealth, that cast members possess. By "complex," I mean that more of these properties are heterogeneous. By "dynamic," I mean that these are changing, rather than remaining constant.

Greater complexity in identity (e.g., ethnicity, professional origin, or even interests and objectives), in relationships (e.g., tie content involving various combinations of such ties as friendship, advice, professional), and in resources poses an "action problem" to anyone operating in such a network context. In other words, it presents a challenge of coordinating people with different interests, unique perspectives, and language (Obstfeld 2005). Increases in complexity demand more skilled brokerage activity in order to produce cooperation, coordination, or other results. While broker facilitation occurs in all structures, such facilitation becomes more labor-intensive in the face of complexity and dynamism, where the broker needs to do more active coordinative and translation work (Beckman and Haunschild 2002).

In increasingly complex, dynamic contexts found in project-based organizing, the value of generating and facilitating collaboration among diverse parties and the diverse resources and information they bear is greatly enhanced.

Tertius iungens brokerage (no doubt in concert with the conduit and *gaudens*) becomes a driving force in conceiving, initiating, and facilitating novel collaborations over time. The simple ability to engineer one optimal collaborative arrangement in the face of uncertainty represents a valuable contribution by itself. *Iungens* behavior, however, also suggests the ability to produce sequences of *tertius* activity over time as the creative project evolves. Furthermore, *iungens* combinatorial action provides a direct means for generating and testing more combinations in the face of increasingly complex, dynamic contexts, enhancing the potential to see opportunities to connect complementary, rather than redundant, alter attributes such as resources and abilities.

By connecting those with differing ties or attributes, *iungens* brokerage brings with it the corresponding challenge of coordinating dissimilar backgrounds and interests, described earlier as the "action problem" (Obstfeld 2005). Such complexity presents the greater potential broker payoffs associated with novel combinations but also presents a greater risk of incompatibility and therefore failure. Stated differently, *tertius iungens* behavior positions actors to generate the requisite variety of novel combinations that better match the conditions found in increasingly complex, dynamic contexts. By contrast, under conditions of homogeneity and stability, fewer combinatorial opportunities are possible, rendering *iungens* brokerage less valuable and *gaudens* activity, which may restrict access, more valuable.

Conclusion

As we seek to test the utility of the framework and ideas put forth in this chapter, it is clear that important empirical challenges remain. First and foremost is the need to measure all three brokerage orientations alongside social network structure. One measure for the *tertius iungens* already exists (Obstfeld 2005). Recent work (Grosser et al. 2015; Soda, Tortoriello, and Ioriao 2015) has begun to develop corresponding measures for *tertius gaudens* and conduit brokerage. Ideally, any such empirical approach would also ascertain the mix of brokerage behaviors that actors employ over time, rather than simply characterizing a given actor as exemplifying one brokerage behavior or another. Obviously, longitudinal data represent a more significant methodological challenge. (See Quintane and Carnabuci, forthcoming, for an example of such an approach using e-mail data.)

New empirical approaches must also take stock of the context in which those brokerage behaviors are measured. In one of the first more sophisticated empirical studies to capture both network structure and brokerage process, Soda et al. (2015) compare the performance of *tertius gaudens* ("arbitraging broker") and *tertius iungens* ("collaborating broker") employees and find that *tertius gaudens* behavior increases the performance of those holding a brokerage (structural hole) position in a human resources (HR) department. The department was tasked with sharing best practices throughout an internationally networked organization, rather than innovating collaboratively. The authors interpret the findings to suggest that HR employees with a *gaudens* orientation do better, because they gain more exclusive access to other units' best practices (as opposed to creating novel combinations). One could imagine an alternative setting involving an HR department within a fairly stable organization and a correspondingly static network with a limited set of opportunities. In such an alternative context, a *gaudens*-oriented strategy that in effect hoards opportunities might also be advantageous. It would be less likely that a predominantly *gaudens* orientation would yield such advantages in the more dynamic, evolving contexts explored above. In any event, the emergence of new measurement approaches introduces enormous opportunities to explore these issues with empirical rigor.

In summary, I have laid out (after Obstfeld et al. 2014) a way to conceptualize a process alternative for social network theory and brokerage based on three brokerage orientations —conduit, *tertius gaudens*, and *tertius iungens*—and argued that skill in employing *tertius iungens* should be seen as the master skill most crucial for getting new things done. A broker's ability to coordinate the involvement of others requires selectively orchestrating the use of all three brokerage orientations. Social skill, or the ability to induce cooperation, is a function of the skilled use of these three brokerage orientations in various patterns—simultaneous, parallel, and sequential—appropriate to the situation at hand.

In the next chapter, I turn to an in-depth treatment of knowledge articulation: the social process by which knowledge is made more explicit, useful, or relevant to the situation at hand, and which occurs alongside brokerage process as a means for getting new things done.

2

Knowledge Articulation

While the social network coordination inherent in the brokerage processes examined in Chapter 1 is a key contributor to getting new things done, it can't be considered apart from the process by which knowledge is moved, translated, and put to use in those relationships. For brokers to get new things done, they not only must coordinate linkages but must also engage with the actors they are coordinating to share, reshape, recombine, and in some cases restrict the flow of knowledge in order to, in the words of Harrison White (2008), "get action."

Brokers establish common ground and buy-in directed toward innovation-oriented, coordinated action through the effective use of knowledge. I refer to the action-oriented use of knowledge as *knowledge articulation* which I define as the social process by which knowledge is made more explicit, useful, and relevant to the situation at hand (Obstfeld 2001; Weick, Sutcliffe, and Obstfeld 2005). This definition pertains to the active pursuit of shared understanding, purposeful communicative exchange, or advocacy, primarily through talk but also through other representational forms (Star and Griesemer 1989; Taylor and Van Every 2000; Carlile 2002) in order to distinguish it from a more general category of communicative behavior.

The inextricable pairing of brokerage and knowledge processes is reflected in two different meanings for the word "articulate." One meaning of "articulate," "to join or unite," is consistent with my treatment of *tertius iungens* brokerage in Chapter 1.[1] A second meaning of "articulate," "to express distinctly,"[2] corresponds with the use of knowledge articulation by Winter (1987; Zollo and Winter 2002) and others (Obstfeld 2001, 2012; Tsoukas 2005; Dougherty 2004), and connects to an intellectual tradition in knowledge-based theories of organizing (e.g., Grant 1996; Spender 1996; Teece 1998), and to Polanyi's (1958) distinction between tacit and explicit knowledge. It is this second usage I turn to now.[3]

Though network- and knowledge-based action necessarily co-occur, I will treat network and network process as separate and distinct from knowledge and knowledge process. Network ties are defined by specific content (e.g., advice, social support, information exchange) and are influenced, managed, or facilitated through brokerage processes. Knowledge, on the other hand, reflects the stock of experience and understanding held by an individual, and knowledge processes concern the communicative behaviors through which a strategic actor puts that knowledge to use in the pursuit of action.

While the network literature has long recognized the connection between networks and knowledge, the predominant conceptualization of this connection is the tie as a "pipe" (Podolny 2001) through which knowledge flows. From this perspective, it follows that an actor's knowledge is influenced by the network "plumbing" within which she or he is embedded (e.g., open or closed networks) as well as by nonnetwork measures located "in the node," such as education, experience in the firm, or social or technical knowledge (Obstfeld 2005). The dyadic relational tie—the building block of social network analysis—also has a dynamic communicative dimension in that it serves as a key locus of knowledge exchange.[4] From this perspective, one's stock of knowledge is relational; it concerns both what one carries to a communicative exchange as well as how that stock relates to those with whom one engages. This relational aspect of knowledge takes into account the extent to which two or more actors share common knowledge (Moldoveanu and Baum 2014; Carlile 2004) as well as the nature of the knowledge exchanged and the project toward which that knowledge is directed.

The shared understanding that underpins new action, in turn, can be characterized as mutual intelligibility, which is "achieved on each occasion of interaction with reference to the situation at hand" (Suchman 1987, 50–51). This emphasis on mutual intelligibility moves to the foreground the dynamic, social, intersubjective nature of knowledge processes, and moves to the background the abstract, decontextualized representations often associated with organizational knowledge (Tsoukas and Mylonopoulos 2004). By "dynamic, social, and intersubjective," I mean the kinds of communicative processes that may never evolve into anything resembling formal representations or codified knowledge, but which nevertheless critically facilitate the movement, creation, and application of knowledge. As such, my analysis of knowledge articulation shifts attention away from *representations* in favor of *representing*, and the means by which representing influences action.

This chapter examines the essential role of knowledge articulation in enabling brokers to mobilize and coordinate others' actions to get new things done. The chapter is organized as follows. First, I examine the tacit/explicit conceptualization of knowledge and its implications for knowledge articulation. Second, I revisit Carlile's 3T model (transfer, translation, and transformation) and how it bridges between brokerage and knowledge articulation (2004). Third, I focus on three initially dyadic processes for articulating knowledge: mutual intelligibility, persuasion, and enlistment. Then I turn to ethnographic data to illustrate knowledge articulation in terms of five practices or communicative dimensions: moving between back stage and front stage; moving between complex and simple; moving among the past, present, and future; balancing familiarizing and defamiliarizing; and establishing credibility by laying down markers. Finally, I return to the relationship between brokerage processes and knowledge articulation in getting new things done.

Putting Knowledge to Use
The Middle Range between Tacit and Explicit Knowledge

By defining knowledge articulation as the social process by which knowledge is made more explicit, useful, or relevant to the situation at hand, I bring to the surface the distinction between *tacit* and *explicit* knowledge (Polanyi 1958). Most definitions of tacit knowledge emphasize relatively unconscious, automatic, taken-for-granted understandings that are difficult to surface or communicate (e.g., Collins 1985; Nonaka and Takeuchi 1995; Spender 1996; Gourlay 2004). Tacit knowing is often understood as automatic knowing (Bargh 1989; Hasher and Zacks 1984) that is involuntary, effortless, and occurs outside of awareness (Reger and Palmer 1996), as with the frequently cited example of bicycle riding (Polanyi 1958). Polanyi's famous observation, "We know more than we can tell" (1966, 4), conveys the difficulty and often the impossibility of communicating tacit knowledge (Von Hippel 1994; Nonaka and Takeuchi 1995; Kogut and Zander 1996; Brown and Duguid 2001). Explicit knowledge, by contrast, is relatively conscious and lends itself to codification. Not surprisingly, this makes explicit knowledge easier to communicate and transfer (Grant 1996; Spender 1996).

The relationship between tacit and explicit knowledge has been the subject of considerable debate. Nonaka and Takeuchi (1995), for example, presented the "conversion" of knowledge from its tacit to its explicit form as "knowledge creation" and, by extension, as the engine of firm innovation; others subsequently pointed to the problems inherent in this characterization. Subsequent work emphasizing knowledge movement (e.g., Zollo and Winter 2002; Dougherty 2004; Weick et al. 2005; Tsoukas and Vladimorou 2001; Carlile and Rebentisch 2003; Tsoukas 2005) largely skirts discussion of the boundaries of tacit and explicit knowledge and of the relationship between the two (Alvesson and Kärreman 2001; Gourlay 2004), emphasizing instead how such knowledge is used (Tsoukas 2005).[5] The idea of making knowledge more explicit is consistent with the concept of a continuum of tacitness (Inkpen and Dinur 1998) that ranges from the tacit and inarticulate, on one end, to the highly codified on the other. Somewhere in between is a "middle ground" where knowledge emerges from tacitness but is not yet expressed in any codified form (Boisot 1995). Such a continuum suggests a continual, though partial and imperfect, surfacing of knowledge.

Mounting evidence in both cognitive psychology and social cognition supports the characterization of individual knowledge as occurring on such a continuum, and rejects the once widely held understanding of automatic processing as an "all or nothing" proposition (Bargh 1989; Reger and Palmer 1996). If we consider individual or shared knowledge as the proverbial iceberg, the larger underwater portion of the iceberg would correspond with the relatively larger size of an individual's tacit stock of knowledge as compared to the smaller above-water portion associated with an explicit stock of knowledge. In between we would find a middle range of conscious but uncodified knowledge (Boisot 1995) that I call *emergent explicit knowledge* (Figure 2.1). This middle range is sometimes overlooked by frameworks that posit tacit and explicit knowledge as a simple binary distinction, or fail to address the transitional states characteristic of many knowledge processes. This middle ground, I argue, is where new knowledge and action emerge, in the form of knowledge articulation.

Polanyi (1958) used the term "articulation" but never defined it. Others have employed the articulation concept to reference the processes by which knowledge is made more explicit (Obstfeld 2001; Winter 1987; Benner 1994; Spinosa, Flores, and Dreyfus 1997; Zollo and Winter 2002; Dougherty 2004, 2006). The knowledge processes associated with many forms of tacit knowledge are not salient because the action associated with tacit knowing—whether

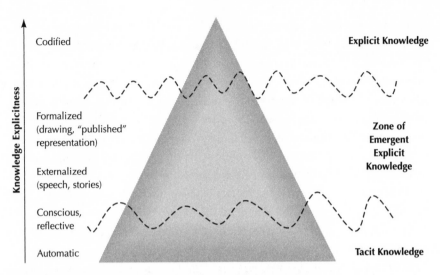

Figure 2.1. Relationship of Tacit and Explicit Knowledge
SOURCE: Original figure.

manifested in the form of skilled individual action or shared in the form of routines (Nelson and Winter 1982)—is typically more automatic (Boisot 1995). Similarly, both tacit and highly codified knowledge, for different reasons, carry a strong, implicit legitimacy (Rogers 2003; Boisot 1995; Brown and Duguid 2001) that reduces the need for justification.[6]

I argue, therefore, that knowledge processes are most salient and relevant for mobilizing action in a middle range of emergent explicit knowledge, somewhere between fully tacit and explicit knowledge, where the lifting of knowledge out of a more tacit state is driven by a new need to share some understanding, move it across communities, persuade others, or change or codify a practice in order to facilitate its more widespread or consistent usage. Active, fleeting representational activity in this middle range between the tacit and explicit is often necessary to generate the joint coordinative action necessary for product and process innovation. This active orientation toward shared understanding is imperative, given the change in thinking, behavior, and even social order that is often needed to mobilize support for innovation.[7]

Making Knowledge Relevant through Translation and Transformation

As introduced briefly in Chapter 1, Carlile's 3T framework (2004) suggests that knowledge can be shared through *transfer*, *translation*, or *transformation*, depending on the extent of the difference in shared meanings across the boundary being crossed. Although Carlile's framework addressed within-firm product development knowledge exchanges, it applies more broadly to cross-boundary knowledge processes occurring across organizations and networks. Also, although Carlile's 3T model is commonly applied to capture the differences in dyadic exchange—e.g., an exchange between two people representing marketing and manufacturing, respectively—any such dyadic exchange at the boundary typically often involves subsequent efforts by one or both of those parties to coordinate action from within or outside each one's respective communities that are implicated by that exchange. So while the cross-boundary knowledge exchanges I discuss below focus on dyads, it is important to recognize that invisible triads sit behind the dyadic exchange and are activated as actors try to elicit and mobilize support. If, for example, we return to Steve Jobs's visit to Xerox PARC, we might first define the nominally "dyadic" exchange between Jobs and the Xerox PARC technologist as the salient knowledge exchange. If we look more broadly, however, we would see that dyadic exchange as situated within two triads, first the triad formed by Jobs, the Xerox PARC technologist, and his Xerox colleagues whose research he had been deputized to represent, and later within the triad formed by the Xerox technologist, Jobs, and the Apple staff member that Jobs deputize to develop the mouse.

I refer to Carlile's three approaches as different ways of "rendering" knowledge, in the sense that rendering can be alternatively understood as "transmitting," "reproducing," "delivering," "representing," "processing," "melting down," or "giving birth to"—a sequence of meanings that suggests a progression from transfer to translation to transformation.[8] Knowledge transfer is critical to mobilizing different actors, but in and of itself it is insufficient to the task. As strategic actors work to forge mutual intelligibility, feasibility, and ultimately enlistment, knowledge translation and transformation become more important. In the following pages, therefore, I revisit Carlile's 3T model (2004) in greater depth, with an emphasis on translation and transformation as the two facets of knowledge articulation.

Knowledge "transfer" is the most straightforward knowledge-movement challenge, in which (1) the differences between actors are known, (2) a common lexicon exists that is sufficient to move knowledge across a boundary without substantial change or manipulation (Carlile 2004), and (3) minimal reformatting of knowledge is required. In the classic act of knowledge transfer, an actor detects an idea or practice in one area that can be moved to another.

How do we tease apart the roles of social network structure, conduit brokerage, and knowledge articulation? A broker's network exposes him or her to certain ideas, social facts, gossip, or opportunities. That exposure is a function of social network structure, whether open or closed, although per structural holes theory an open network is more likely to expose the broker to new ideas (Burt 1992). Exposure to various new ideas or opportunities, however, still leaves in question whether the broker has *identified* potentially transferable ideas and opportunities, and subsequently, whether he or she *acts* to move them. If this sequence of exposure, identification, and movement involves primarily the transfer of opportunities and ideas "as is"—that is, with minimal translation—it implicates conduit brokerage, but little if any knowledge articulation.[9] The combination of network structure (which may afford the broker a unique vantage) and conduit/transfer (the choice to identify and move something) constitutes an active response to an information advantage. The broker in this case engages in opportunity *curation*—that is, an act of "analyzing and sorting . . . and presenting it in a meaningful and organized way."[10] Knowledge curation, however, which I will return to later, implies translation.

"Translation" (Czarniawska and Joerges 1996; Latour 1987) involves communication across a semantic boundary where different meanings or interpretations exist. Translation is a semantic adjustment of a communication, in anticipation of or response to the attitude or understanding of the other (Mead 1934). I might translate an idea in anticipation of your expertise, ignorance, or preferences. Alternatively, I might translate in response to your previous comment because it reveals a particular preference or perhaps a misunderstanding. Finally, translation may occur simply to establish and maintain mutual intelligibility, establish possibility, or ultimately achieve feasibility and buy-in. "Transformation" can be defined as a significant rearrangement, new representation, resynthesis, or new synthesis of knowledge, often in response to the challenge presented by significant novelty or gaps in understanding.[11] Transformation is required where incompatible interests prevent the establishment of shared meaning. Such incompatible interests are lodged in the knowledge and practices associated with two or more conflicting, interdependent areas, and so are inherently political. In the absence of resolution through the imposition of overwhelming power, common interests need to be developed as a means of moving forward.

Transformation, according to Carlile (2004), involves the substantive mutual altering of knowledge through representing, specifying, negotiating, or compromising, to accommodate the differences in knowledge present at a boundary. The signature of transformation is that one or more parties change their position or understanding to adopt an innovative course of action.

Knowledge Articulation Goals:
From Mutual Intelligibility to Enlistment

In the purposeful articulation of knowledge, in pursuit of getting new things done, advocates (often brokers) seek to influence shared knowledge (*mutual intelligibility*), views (*persuasion*), or behavior (*enlistment*) (Table 2.1). It is often but not always the case that mutual intelligibility precedes persuasion and enlistment. Let's consider how knowledge articulation helps to get new things done where the "new thing" involves some form of new technology that necessitates some change to an organization's product and/or process.

Table 2.1. Linking Knowledge Articulation and Brokerage

Emergent Goal Progression in Pursuit of Innovation	Process Building Blocks in Action (Social Skill)	
	Communicative Dimensions of Action	Brokerage Orientations to Action
Goal 1: **Mutual Intelligibility** (Shared but sometimes asymmetric understanding)	1. Back Stage ⟷ Front Stage	Conduit
	2. Complex ⟷ Simple	
		Tertius iungens
	3. Past ⟷ Present ⟷ Future	
Goal 2: **Persuasion** (Purposeful pursuit of agreement around possibility and feasibility)		
	4. Familiarize ⟷ Defamiliarize	*Tertius gaudens*
	5. Laying Down Markers	
Goal 3: **Enlistment** (Enrolling actors; creating and managing collaborative arrangements)		

Source: Original table.

Mutual Intelligibility

The first goal level of knowledge articulation reflects the extent to which two actors successfully achieve some form of mutual intelligibility (Suchman 1987). Over time, such an accomplishment might appear simple, even automatic, but in fact it may well constitute a considerable communicative accomplishment, one that actors frequently fail to achieve in any meaningful way.[12] To develop such a mutual understanding requires some form of Mead's (1934) symbolic interactionist idea of taking the attitude of the other. In other words, to successfully communicate knowledge in a given exchange, the broker must take the alter, her attitude, and existing state of knowledge into account *throughout* the exchange, in order to assess her understanding and adjust communication as necessary.

Mutual intelligibility can range from a roughly equal engagement, with back-and-forth exchange of meaning, to a more asymmetric level of interest and attentiveness. In the latter case, the broker may be far more solicitous of the alter's interests before, during, and after the interaction than the alter's interest in the broker, perhaps because the broker has a strong investment in gaining the new technology's adoption for which the alter's support is critical, and the alter is indifferent to the technology's adoption.

Persuasion

A second goal of knowledge articulation in getting new things done concerns communicative behaviors in which the actors engage to generate interest in the new technology. Such efforts involve engaging one's interlocutor in the new technology's *feasibility* (i.e., does the new technology have the potential to influence our performance in the way that has been suggested?) and *opportunity* (i.e., is the new technology a good idea given our respective interests?). At this second level, the intersubjective challenge concerns the relative agreement between respective evaluations of the new technology. This is where actors—both advocates and stakeholders—individually and collectively consider the practical consequences of the new technology in light of their personal, professional, and organizational interests.

Each group may have a solid grasp of the other's concerns and interests about what a new technology entails (mutual intelligibility), and may be able to iterate to an altered version of the new technology that everyone agrees can produce the (revised) outcomes that are claimed for it (feasibility), but still disagree as to whether it's a good idea for the organization to adopt that technology (opportunity). In an additional wrinkle on the evaluation of opportunity, several might even agree that while it's good for the larger organization, it threatens to eliminate the need for their department, and they therefore oppose the new technology.

Enlistment

A third goal of knowledge articulation concerns *enlistment* or *buy-in*, the extent to which a broker successfully enrolls an actor,[13] and *mobilization*, the subsequent efforts to knit that support into a broader coalition. The transition from persuasion to enlistment requires a transition from a "merely" favorable attitude toward the new technology to a more active form of support. This enrollment activity has a general resemblance to *tertius iungens* brokerage: the use of knowledge articulation to connect and coordinate previously uncoordinated actors, as noted in Chapter 1.[14]

The knowledge-based emphasis presented here should not obscure the fact that the move from recruitment through enlistment is infused with political calculation. The political aspects surrounding knowledge processes were overlooked in some earlier knowledge-based accounts of innovation, which stressed issues like the tacitness or stickiness of knowledge processes (Nonaka and Takeuchi 1995; Brown and Duguid 2001), but have been recognized by others (e.g., Carlile 2002, 2004; Latour 1987; McCloskey 1998). A new technology typically engages multiple actors at differing levels of support and engagement and with various motivations. Some actors may fully support the new technology, while other actors, being recruited to the existing cluster of support, may view the new technology with some mix of skepticism, calculation, and guarded interest. Irrespective of an individual's position of power, it's rare for any single actor to be able to simply compel the adoption of that new technology. Rather, adoption will require progressive enlistment and mobilization of superior, lateral, or subordinate stakeholders within and outside the organization.

Political alignment and enlistment may be pursued sequentially after mutual intelligibility is established, but enlistment might depend on episodic or sustained mutual intelligibility, and is often pursued in the absence of robust mutual understanding (and may fail or succeed as a result). And as noted, the tenacity with which an actor pursues mutual intelligibility, recruits, and enlists is often asymmetric, given that some advocates may have a far greater stake at getting new things done than others, and as a consequence pay far more attention to the knowledge and interests of their interlocutors.

An important thread of research (Bechky 2003; Carlile 2002, 2004; Tsoukas 2009) has developed an understanding of how the knowledge integration associated with novelty and innovation is accomplished by roughly equally and intensively engaged collaborators who "identify, elaborate, and then explicitly confront the differences and dependencies across knowledge boundaries" (Majchrzak, More, and Faraj 2012, 951). This research describes the in-depth dialogue that leads to the knowledge transformation described by Carlile (2002) as "deep-knowledge dialogue," or what I will refer to as DKD (Majchrzak et al. 2012). DKD involves sustained dialogue through the questioning and probing of collaborators' assumptions in the search for new, jointly produced knowledge that transcends conflicts in local knowledge. It may sometimes require complete breakdowns in coordination and collaboration to drive such engagement as a means for going forward.

Recent work, however, has pointed to the costs and challenges of DKD (Edmondson and Nembhard 2009; Hansen 1999), and suggests that constructive knowledge integration can proceed in the face of novelty without DKD through alternative forms of knowledge sharing (Kellogg, Orlikowski, and Yates 2006; Majchrzak et al. 2012; Ewenstein and Whyte 2009; Schmickl and Kieser 2008). I will refer to a broad swath of such alternative knowledge integration processes—characterized by variable levels of participation less complete than DKD, and often involving experimentation with emergent collaborative technologies—as *provisional knowledge dialogue* (PKD). In the face of high speed, uncertainty, and rapid change, Kellogg et al. (2006) talk about provisional settlements as a means by which projects move forward. Kellogg et al. (2006, 34), citing Girard and Stark (2002), define provisional settlements where team members postpone extensive consensus building on an issue and negotiate a temporary agreement to keep the work going, with the mutual understanding that these areas will be revisited and resolved further down the line. These provisional settlements allow team members, according to Kellogg et al., "to interact on projects without making deep commitments to shared meanings or transformed knowledge" (2006, 40).

Note the chicken-and-egg relationship between the provisional settlement and participation. Whereas the provisional settlement addressed impasses and incommensurabilities that emerged in the loosely coupled, fast-paced setting observed by Kellogg et al., other provisional arrangements might serve not as the means for resolving impasses but rather as the basis for attracting participation a priori. For example, Fred (in Chapter 1) used the promise of an attractive future project as a basis for bringing Gloria and Joann first to meet and then to collaborate. I adapt the broader term of *provisional arrangements* to address both the prospective and reactive connection of parties in the *tertius iungens* spirit. I argue that social skill and, more specifically, knowledge articulation is integral to the pursuit of getting new things done whether through social influence, PKD, or DKD.

Finally, knowledge articulation—whether in the form of mutual intelligibility, the exploration of a new technology's possibility and feasibility, or the pursuit of enlistment and buy-in—is ultimately pursued through varying levels of DKD, PKD, and social influence. Sometimes a socially skilled actor will productively employ social influence along with PKD to move a new technology forward, up to the point at which he or she encounters difficulties that cannot be bridged without a more comprehensive resolution of differences and dependencies constituted by DKD. In the fluid, knowledge-infused social processes associated with getting new things done at NewCar (see below) and its parent company, AllCar (see Introduction), knowledge articulation was the means by which social influence, PKD, and DKD were pursued.

Illustrations of Knowledge Articulation at NewCar

To illustrate socially situated knowledge processes in which knowledge articulation occurs, I turn to field observations from NewCar. In both product and process innovation efforts, NewCar actors drew on their *technical* and *social* knowledge to develop and sell design alternatives, continually responding to shifts in the organizational environment and individual support. At NewCar, the technical dimension of knowledge comprised the skill and experience generated from participation in multiple product development efforts over time. This knowledge derived from the individual's day-to-day experience of designing parts and subassemblies, prior successful and failed product development efforts, and experience gained applying different product technologies and materials. This technical knowledge sometimes meant having a grasp of an existing part or subassembly, of the product's limitations, or of the feasibility of possible changes.

The social dimension of knowledge included the shared, day-to-day experience of designing parts alongside other workers; an understanding of the history, evolution, and interests of different functional areas; the nature of the relationships among these different functions, including points of friction and alignment among the people within those functions over time; the concerns of individuals regarding different product development efforts; and the tactics and timing appropriate for orchestrating divergent or aligned interests at different levels within the organization.

The two vignettes below involve simple dyadic exchanges, although I point out in each case how those dyadic exchanges mask closely related triadic processes through which other actors and departments may be engaged. For methods, see Appendix 1.

Vignette 1: Routine Design Exchange

Rick, the lead stylist (from the styling division) responsible for the aesthetic appearance of the G5's interior, is meeting in a prototyping area with Tom, a designer, and Bob, an engineer, both from the engineering group responsible for executing those designs. The purpose of the meeting is for Tom and Bob to review their most recent prototype mock-ups with Rick to determine if they have sufficiently captured his design intent. In this exchange, Rick advocates for several changes to the prototype door and armrest, some of which involve changes to the digital design itself, and others to the way the part would be manufactured. As he talks, Rick refers to and repeatedly touches a detached armrest prototype while commenting on its plastic surface.

> RICK: You got my door?
> TOM: It's just a [mock-up]. . . . You get two more tools to get it right.[15]
> RICK (*pointing to one part of the armrest*): Could this be thicker there?
> [*Tom asks if Rick wants to see the armrest inserted into the prototype door. The three walk over to a door next to Tom's desk and pop the arm into the door. On the way over there is some brief joking about how the key to working together involves being tough and nice, after which Rick acts out the joke, saying with mock curtness:*]
> RICK: Just do it. . . . [*followed by a facetiously exaggerated friendly response:*] Thank you. . . . [*turning his attention to a small switch surface at the intersection of the door and the handle*] I was hoping this would be lighter. . . . [*then, after inspecting the overall handle:*] This is gorgeous.
> ME (*asking why he was surprised by the quality of the handle's appearance*): Didn't you design it?
> RICK: Yeah, it's my design, but it's more about how the designs get filtered. To see it so close to intended is exceptional.
> JOE (*with surprising sincerity*): Well, I do admire your work very much.
> RICK (*quickly responding*): Ass kisser. . . . [*then, regarding the armrest's surface:*] Make it a good stipple.[16] A fairly heavy stipple. This feels like a Barbie doll stipple. My only concern is that it slips. [*holding the detached armrest and commenting on its plastic surface:*] It's disconcerting. You should make it pretty tight, like a scuba outfit.

This episode features many dyadic exchanges, and illustrates their central role in getting new work done. Rick, although a stylist, has moderately strong ties with many of the engineers with whom he periodically works across the styling-engineering divisional boundary, although such design work at NewCar constantly required the formation of new collaborative clusters involving both new and reengaged relationships. Within the larger exchange from which the above data were excerpted, network dimensions were also in play, with structural holes and conduit brokerage implied by the stylist's relaying of directives and gossip from the styling division, to which the engineering staff had limited exposure, and reciprocal information from the engineering staff back to the stylist.

Knowledge articulation in the episode involves the stylist's efforts to convey his design objectives in order to realize his aesthetic vision in the form of the physical prototype and ultimately the production vehicle itself, and the engineers' rejoinder regarding what is feasible. The relatively straightforward exchange represented one in a series of ongoing interactions in which Rick had to convey styling objectives not fully captured in the formal three-dimensional design and other accompanying sketches.

The pace of the exchange, along with the G5 stylist's expression of satisfaction with the prototype armrest (indicated by his comment, "To see it so close to intended is exceptional"), suggests a high degree of mutual intelligibility. The engineering group's respect for Rick's work reflected in Bob's expression of admiration suggests an additional advantage Rick can leverage in gaining agreement around possibility, feasibility, and enlistment. Finally, Rick employs different articulation "devices" (e.g., analogy, metaphor, physical objects) to convey his design objectives in order to achieve his overall aesthetic vision for the G5 interior.[17] In some cases, unembellished verbal communication is enough (e.g., "Could this be thicker?" and "I was hoping this would be lighter"). Here, however, the reference to the physical prototype part plays a crucial role. In the latter part of the exchange, the use of metaphor and analogy (e.g., avoiding a Barbie doll stipple, "pretty tight like a scuba outfit") was an important part of the knowledge articulation process.

Vignette 2: Advocating for Paper Car Database

The next exchange is excerpted from a longer hallway exchange, which takes place at a higher administrative level involving more process (versus product) concerns and more attention to politics and enlistment.

As the manager of Body-in-White (an area within the "Body" division), Brad is responsible for working out a precise stamping of the sheet metal exterior of the car, widely considered to be one of the most technically demanding and important automotive manufacturing tasks. As a result, Body was seen as one of the most important, influential, and high-status engineering units. Brad is also the lead manager for developing and executing the "Virtual Build"—a cutting-edge digital database—and the associated software, "Virtual Car," containing geometrically accurate 3-D representations of every car part, and capable of simulating and testing how those parts fit together.[18]

In the following exchange, Dan seeks to convince Brad (and by extension Brad's department) to actively support the use of an older comprehensive parts database called the "Paper Car," crucial to coordinating the parts production essential to multiple prototype builds and, ultimately, full-scale production. Though Brad and Dan come from two different engineering divisions (body and program management, respectively), they have a history of collaboration and friendship.

> BRAD: The "Paper Car" is a waste. The Virtual Build is all that counts. I didn't know how to tell that to Pete [*the senior executive program manager and Dan's boss*] in the [weekly senior executive meeting].
>
> DAN: They are not the same thing. We are not a car company. We are an information company. We shit out cars. . . . But we are an information company. We have all this data about the car [*motioning with his hands about three feet above the ground to suggest a small bucket or pool*] and from that we create the Virtual Car over here and the Paper Car over here [*miming the existence of a second and third bucket*]. The [Virtual Car] is to verify the integrity of the data with [the] geometry [of the parts]. Now we are doing it in a prototype stage. The Paper Car is a validation of your car in the prototype build specification database.
>
> BRAD: I accept that.

The impromptu hallway discussion from which this exchange was taken included small talk, good-natured teasing, and attention to issues central to Brad and Dan's shared responsibility for the G5's design and manufacture. In such exchanges, abstract, longer-term issues were subordinated to the pressing problems at hand. Brad's candid skepticism about the Paper Car database and admission regarding his reluctance to tell Dan's boss about that skepticism suggest a fair level of trust between the two men. Given his responsibility for Body-in-White and Virtual Car, Brad approached the conversation with a typical body department desire to avoid more pedestrian operating issues and a corresponding view of the Paper Car database as an annoyance with little added value.

Given his overall responsibility for coordinating the G5's design and production, Dan approached the conversation with a conscious desire to get Brad and his department to support and maintain the Paper Car database. As the G5's program manager, Dan shared Brad's commitment to the new technology implementation but also was responsible for ensuring another critical production imperative: the availability of parts necessary for an accelerated production of G5 prototype vehicles crucial to the ultimate success of the G5.

The individual but socially situated perspectives that Brad and Dan brought to their hallway exchange reflect a variance in perception of feasibility and, as a result, opposing positions with respect to the Paper Car database. Dan reframed NewCar as an "information company" in an attempt to persuade Brad to support the Paper Car database. He was mindful that a successful prototyping and production effort required maintaining the Paper Car database to coordinate prototype builds, shipping, and manufacturing while successfully developing the Virtual Car software and database with which Brad was preoccupied—to coordinate prototype builds, shipping, and manufacturing.

In the hallway exchange, Dan's persuasive efforts involve a balance of confrontation and appeal. He rejects Brad's initial contention that Paper Car is a waste, but follows with an unexpected framing that Brad abandon an outmoded perspective of AllCar as a car company and instead adopt a new understanding of AllCar as an information company. Such a characterization of NewCar would appear to have additional appeal for Brad, given the information technology focus associated with his Virtual Build and Virtual Car responsibilities. Dan was also endeavoring to recalibrate Brad's focus on the Virtual Build to a broader perspective associated with the G5's prototyping and production objectives. His remarks confronted Brad, but they also offered an alternative, broader common meaning that invited Brad to share his concern for the success of the G5.

The informal hallway exchange was a dyadic interaction, but as with the previous vignette it also involved many hidden triadic relationships that included the two managers, the executive engineers to whom they reported, their peers, and the staff in their respective organizations. One triad concerned the relationships among Brad, Dan, and Dan's boss (referenced in the exchange). Other implicit triads involved Dan, Brad, and each of the other senior managers Dan hoped to persuade to use Paper Car by enlisting Brad. Their dyadic exchange also involved queries from each about the work of others within their respective organizations. Such exchanges bore directly on how they coordinated incremental or more substantive changes to the ongoing G5 design routines for which they were responsible. Dan described this *iungens* orchestration among functional areas like Body and his department several months later:

I created relationships with my direct reports and functional [areas]. I create
links between my reports and . . . managers [*from other G5 areas*] to [work to-
gether]. I work on both sides of that. Get my people comfortable with that and
get managers comfortable. When a ball comes off the [court] they're comfortable going to [my] program management person.

Dan's exchange with Brad, therefore, hinged not only on mutual intelligibility,
persuasion, and enlistment within their dyad, but multiple downstream patterns of conduit, *tertius iungens*, and *tertius gaudens* brokerage radiating outward
from their exchange regarding the Paper Car database's feasibility, and influencing downstream enlistment and mobilization. A decision by Brad to support the Paper Car database would (and did) initiate multiple triadic patterns
of communications between his Body engineering staff and Dan's program
management staff, as well as the purchasing, prototype production, and operations departments, in order to make the appropriate operating changes to their
ongoing design work.

Five Communicative Dimensions of Action

Having illustrated two relatively brief episodes of knowledge articulation to get
new things done, I now turn to five fundamental communicative dimensions
along which actors articulate knowledge (sometimes reciprocally and of course
in clusters greater than two): between back and front stage; between complex
and simple; among past, present, and future; balancing familiarizing and defamiliarizing; and laying down markers. (The role these dimensions have in the
brokerage process will also be developed further in several extended ethnographic cases in Chapter 4.)

Moving between Back Stage and Front Stage

Given the multiple intersecting social worlds that actors occupy when communicating to get new things done, they constantly make choices about what
information to put forth, how and when to do it, and alternatively, what information to withhold or leave unsaid. This is a communicative dimension that I
refer to as "moving between back stage and front stage."

The choices actors make along this dimension constitute the social cor-
relate of moving knowledge between a more tacit and a more explicit form.
The dimension involves the surfacing of heretofore unarticulated, sometimes
embodied knowledge, and also curating one's diverse body of socially situated
knowledge to determine the content, framing, and level of depth appropriate

to a given situation. The knowledge that is articulated surfaces privately held ideas for public discussion, making that knowledge more social.

While individual knowledge is socially and culturally constituted (Lave and Wenger 1991), it is ultimately held and expressed (or not expressed) by individuals (Tsoukas and Vladimirou 2001). In this sense, knowledge articulation involves a disembedding or lifting and expressing of selected privately or socially held knowledge. Per the double interact, the lifting and expressing of knowledge are followed by some form of reciprocal communication, where each interlocutor is making similar choices about what to share with the other.

The reference to back stage and front stage draws loosely on Goffman's (1959) classic work on presentation of self, but with important distinctions. In brief, Goffman saw many exchanges in the social world as dyadic performances with relatively fixed characteristics. Of key importance was the structure of the setting that demarcated a front stage, where the performance took place, and the back stage, where performers prepared for that performance. In Goffman's case of a restaurant, the dining room, where the waiter serves food with elegance and sophistication, is front stage, while the kitchen, where the waiter is free to direct a stream of profanity at the cook because the food for his table is not yet ready, is back stage. Goffman saw these performances as relatively fixed, in terms of both how the stage is defined and the highly routinized nature of the performance. I invoke the metaphor to demarcate the actor's ongoing judgments about what knowledge and information to put forth and what to withhold in a given exchange, as well as the depth and framing most appropriate to the situation.

But I take considerable liberty with Goffman's metaphor. First, where Goffman used both front stage and back stage to denote public spaces, I am using back stage to include both a social domain and the interior reflections of an actor as he or she makes choices as to what ideas and what format to employ in a given moment.[19] Second, as noted, where Goffman saw the stage's parameters as essentially fixed, I argue that the stage for knowledge articulation is far more fluid in terms of setting and audience. Third, Goffman assumed the front-stage/back-stage distinction was primarily dyadic, sometimes referring to the performance as involving "two teams" and emphasizing the role of the performer and an audience.[20] Though I have taken dyadic communication as an important point of departure, I argue again that some of the most important knowledge-articulation performances involve three or more parties. In addition, the roles of performer and audience likely rotate among two or more of the actors involved. Fourth, whereas Goffman emphasized the routine nature of performances (e.g., of waiters, executives, factory workers, umpires, housewives),

I stress that *performances are inherently extemporaneous*, in response to shifting constellations of assembled actors whose interests must be differentially addressed or reconciled. But rather than undermining Goffman's front-stage/back-stage metaphor, these qualifications actually underscore the performance-oriented nature by which knowledge articulation gets new things done, especially in less predictable settings.

Drawing on artifacts and speech in his meeting with engineering, Rick expresses his personal sense of how his design should be realized. The digital drawings he provides to engineering constitute an initial stage of knowledge articulation, but can go only so far in expressing his intent. The production of an actual prototype part that can be held and discussed provides important additional means by which mutual intelligibility can be pursued. Through the knowledge articulation that accompanies the artifact, Rick makes further refinements, some of them rather "straightforward" (e.g., "thicker," "lighter") from a discursive standpoint, while other more complex requests depend on the use of metaphor and analogy.

In their unscheduled conversation in a NewCar hallway, Dan and Brad each give voice to their personal and collective experience, understanding, and beliefs in connection with Paper Car. Brad's initial disclosure, made possible by the trust he has with Dan, causes him to reveal an until-then deep "back stage" skepticism about Paper Car—"The Paper Car is a waste"—which in turn sets the stage for Dan's candid response: "We are not a car company. We are an information company." Both Brad and Dan selectively move their more privately held back-stage understanding into the front stage of their exchange, thereby reaching a shared understanding of how to proceed with the G5 design effort. Given their friendship and its associated trust, each is more forthcoming than he might have been in other contexts.

The knowledge being articulated may be held personally or collectively. In the most obvious case of articulation, and the one that most directly corresponds with Polanyi's (1958) original treatment of articulation, an individual articulates his or her own personal knowledge. Alternatively, an individual may articulate the knowledge held by the group or organization to which she belongs. In this second case, the individual draws on her own experience, along with some grasp of the shared knowledge held by the group. This articulation may represent the group's technical knowledge, or its social knowledge in the form of preferences, motivations, or beliefs.

There was a picture posted on many cubicle walls associated with the G5 project, which served as an influential articulation of a shared understanding.

It showed a train, clouds of smoke billowing from its smokestack, rounding a bend in front of tracks that were still being laid (Figure 2.2). The picture was used prominently by managers to articulate to G5 engineers and senior managers the kind of just-in-time work necessary for the G5 to succeed. The drawing was meant to depict both NewCar's innovative past and its current challenge: to design the G5 while implementing major new technology and process initiatives. More informally, the picture was also seen as articulating the risk taking and innovativeness in which many NewCar employees took pride. The train indicated, in the words of one frontline engineer, how NewCar and the G5 was "constantly pushing the envelope." Nine months into my fieldwork, I discovered that Gerard, a key informant, had commissioned the train picture as an expression of NewCar's ability to innovate in tight time frames and with limited resources:

> We had a conversation with a few people and somebody said, "You know, it's like . . . the cart before the horse," and somebody said, "Yeah, it's like trying to build a train while you're still laying the track." . . . And then somebody said, "You know, we should draw that."

Figure 2.2. The NewCar Train Picture
Source: Unofficial figure from defunct automotive firm, 1988.

He further revealed that the picture was specifically prepared as part of the presentation to one of the highest-ranking executives in the company, Rod, to illustrate the challenges that the G5 confronted:

> It was specifically drawn up to be presented to Rod in a technology review. [The train picture] was on a poster board saying here's our challenges and there [were] three challenges in technology. And the big challenge—the last challenge—said, keep the train on the track moving. . . . And we showed him that.

According to Gerard, the senior managers' impression of the train picture was positive, consistent with the NewCar tradition of innovation and risk taking: "The execs [were] there, they all thought it was a wonderful thing. [Their response was,] 'You're right. That's what we do.'"

Moving between Complex and Simple

The contrast between the brevity of the Paper Car exchange and the depth of knowledge that lay behind the two articulated positions illustrates how articulation often involves the reduction of knowledge to fit the situation at hand. Reduction reflects an alternative reading of Polanyi's observation, cited earlier: "We know much more than we can tell" (1966, 4). Given the practical time and attention constraints on social exchanges, interlocutors can only articulate a fraction of what they know; what is communicated is necessarily always incomplete. Reduction, therefore, involves a stripping away of "irrelevancies" to convey only that which is most relevant to the situation and the interlocutor at hand. Reduction involves knowledge curation, as contrasted with the opportunity curation that takes place in conduit brokerage. The use of various articulation devices (e.g., analogies, metaphors, stories, informal sketches, and physical objects) can help strip away the chaff of less-relevant knowledge and construct a communication intended to be useful in the situation at hand.[21]

Knowledge articulation is accomplished through gestures (Mead 1934), symbols, and language (Dewey and Bentley 1949; Boisot 1995; Gourlay 2004) but only rarely results in fully codified knowledge (e.g., a written list of steps or rules). Such knowledge articulation makes some aspect of knowing more manifest (and likely "backgrounds" others).[22] To further develop the idea of knowledge articulation, I build on qualitative studies of organizational knowledge processes (e.g., Lave and Wenger 1991; Dougherty 1992a, 2004; Orr 1996; Orlikowski 2002; Carlile 2002, 2004; Bechky 2003; Stigliani and Ravasi 2012). I empirically examine the complex coordinative networks of action found in an automotive design process, using both quantitative and ethnographic data.

The tension between the complex and simple may involve a more literal focusing of attention (e.g., "like a Barbie doll stipple") or more abstract summaries (e.g., "We are an information company"), anecdotes, or frames.[23] Dan's use of metaphors and illustrative hand motions simplifies a broad set of considerations to an important few while also offering a more complex grasp of NewCar. Prototype vehicle data are never literally found in or transported between "buckets" that can be described with hand motions. Such information is never moved, as his hand motions suggested, like objects. Hand motions, however, in concert with metaphor, simplify complicated, obscure database relationships and fashion a more complex meaning around NewCar being both a car company and an information company.

The move from complex to simple may also involve the iterative fashioning of slogans or tag lines to encapsulate ideas that might facilitate enlistment in innovation initiatives. The G5 program manager repeatedly described the objective of his campaign to create a new prototype parts-purchasing unit as getting "the right parts in the right cars at the right time." Advocates of the broader AllCar Development Process effort, intended to accelerate the design process, adopted the carpenter's maxim "Measure twice, cut once" to communicate the philosophy of rooting out design errors early to prevent subsequent process delays. A grassroots business process reengineering effort recognized the respect commanded by a legendary prototype parts manager, Craig Jones, who embodied the coordinative role they envisioned for a new group they hoped to create. They began informally suggesting that the new group would function "like Craig Jones."

In each of these cases, innovation advocates improvised succinct statements to communicate and attract support for innovation initiatives.[24] Such encapsulations begin to approach the frames that Kaplan (2008) found so effective in mobilizing support for strategic initiatives. In the field, social skill is often in evidence as actors exhibit more or less effectiveness in selecting, winnowing, and refining the ideas and information most likely to generate understanding and support from a given audience.

Moving among the Past, Present, and Future

Mobilizing action to get new things done involves the repeated necessity of bringing knowledge from past experience to bear on present situations, in order to elicit support for joint action in pursuit of desired future states. Knowledge is acquired over time, residing in memories associated with the contexts in which it was acquired. Hargadon and Sutton (1997; Hargadon 2002) have explored, for example, how knowledge brokers link past experiences to current situations to create innovative solutions. Similarly, as NewCar employees engaged new problems and solutions, they brought their understanding forward to address the situation at hand. In his Paper Car discussion with Brad, Dan drew from that past experience both explicitly and implicitly to address the conversation at hand.

Past-to-present knowledge articulation commonly occurred in design conversations. In Rick's discussion about the type of armrest surface he hopes to achieve, he inserts a reference to the Barbie doll stipple and scuba outfit, analogies instinctively plucked from a cultural past that he and his audience shared. The straightforward exchange represented only one in an ongoing series of interactions in which he had to convey various styling objectives to the engineering group, in ways that would gain engineering support for (or at minimum, compliance with) his overall aesthetic vision for the G5's interior.

In another case, the tension between the past, present, and future took the form of a story told by Frank, a NewCar senior vice president, to encourage his direct reports to take more risks. In a G5 meeting with his managers, Frank relates how, as a young NewCar engineer, he went about solving a "power hop" problem in which the vehicle under design jumped or bounced uncontrollably when pulling heavy loads on soft ground:

> It was in the winter and we had a terminal power hop. We were building a [*name of vehicle*] on an [F2] platform. Our four-cylinder [engine] was shaking to death. We were doing cardboard and free body designs in the office and trying to figure out the mechanism. We were halfway to figuring it out. . . . We had a pile of sand. We would [hook] a pickup [truck to the prototype] . . . to have a simulation of going up a hill. We had our own sand pile up front. We were out there with our blue shop coats in the middle of the afternoon. The neighbors were thinking . . . "They keep getting stuck there and having to pull out with that little [car]." It was the most fun I had in twenty-six years. . . . Guys like Harrison Stickel, old-timers, were good at that. Now we spend a bazillion dollars with contractors.

The senior manager's story bracketed and ordered a small subset of impressions from a complex set of events that occurred one winter afternoon nearly twenty years earlier. The story did not exist a priori, but was constructed retrospectively (Weick 1995) at some point in time and then reinvented in each retelling. This story brought forward in time an incident that illustrated and burnished NewCar's "cowboy" culture of risk taking in pursuit of innovation.

This was but one of many informal narratives employed at NewCar to move the past to the present in pursuit of future outcomes.

Balancing Familiarizing and Defamiliarizing

To the extent that knowledge articulation is deployed across a boundary, it is intended to convey knowledge to alter a stakeholder's knowledge, view, or behavior. Such a transforming objective involves a balance of *confronting* and *making an appeal*.

Since the vast majority of organizing is lodged within routines, innovation requires disruption of routinized ways of communicating and getting things done. Nelson and Winter suggest that routines represent a truce between competing interests, and argue that "the state of truce is ordinarily valuable, and a breach of its terms is not to be undertaken lightly" (1982, 111). Efforts to innovate often must disrupt such truces in order to reconfigure actors and meanings. Stated differently, the knowledge associated with ongoing routines is largely tacit, familiar, and invisible. In order to initiate new action, the actors I observed often took steps to make a salient, even jarring, break from existing routines and associated understandings through breaching (i.e., a violation of a social rule or norm) and defamiliarization (Garfinkel 1967; Taylor and Van Every 2000).

I use the term "defamiliarization" (Shklovsky 1990) to refer to advocates' efforts to call into question taken-for-granted (i.e., tacit) understanding. Familiarization, by contrast, involves the opposite: the process of making foreign or strange ideas familiar to the intended audience. Both defamiliarization and familiarization are involved in the articulation processes. The Russian literary theorist Viktor Shklovsky first introduced the idea of defamiliarization in 1917 to describe the capacity of art to overcome the deadening effects of habit and convention (Lodge 2012). Defamiliarization is a translation of the Russian word *ostranenie*, which means, literally, to "make strange" (Rice and Waugh 2001).[25] Art's capacity to break free of habitual, more automatic forms of knowing, as articulated by Shklovsky, was very similar to the capacity that innovation proponents at NewCar displayed for calling into question certain habitual practices that their innovations disrupted.

The Paper Car episode illustrates how knowledge articulation often involves a combination of defamiliarization and familiarization.[26] Dan first defamiliarizes by stating something counterintuitive ("We are not a car company. We are an information company") and then further defamiliarizes by subverting the prized nature of car manufacture, comparing it to a lowly bodily function ("We shit out cars"). This metaphor in effect turns the car manufacturing practice from something being thoughtfully crafted in front of the body, with the hands, to something thoughtlessly excreted from the behind. By disrupting Dan's familiar grasp, he thereby creates the opportunity to significantly alter the orientation of Brad and his department toward the maintenance of the Paper Car database. At the same time, he also makes a familiarizing move by referencing and valuing Brad's existing information technology expertise and using simplifying gestures that translate database maintenance of cumbersome databases to a more tractable idea of moving information with buckets.[27] Schutz captures this familiarizing move in his description of the transition from "strange facts" into "warranted" knowledge:

> If we encounter in our experience something previously unknown and which therefore stands out of the general ordinary order of our knowledge, we begin a process of inquiry. We first define the new fact; we try to catch its meaning; we then transform step by step our general scheme of interpretation of the world in such a way that the strange fact and its meaning becomes compatible and consistent with all the other facts of our experience and their meanings. If we succeed in this endeavor, then that which formerly was a strange fact and a puzzling problem to our mind is transformed into an additional element of our warranted knowledge. We have enlarged and adjusted our stock of experiences. (1944, 507)

Innovation proponents who introduce new ideas are tasked with reducing the innovation's strangeness to their audience. Reducing strangeness is of particular importance when communicating across organizational boundaries or seeking to persuade. Dan's use of metaphor and analogy was initially disruptive, but because they were simple and accessible they also had this familiarizing impact. Ozick indicates, "Metaphor relies on what has been experienced before; it transforms the strange into the familiar ... metaphor uses what we already possess and reduces strangeness" (1989, 280). Rick's reference to Barbie dolls and "tight like a scuba outfit" not only articulates his meaning but also familiarizes his audience with his intended outcome, thus deepening understanding and buy-in. Though both familiarization and defamiliarization can be employed to translate and transform, defamiliarization naturally lends itself to the disruption of taken-for-granted meanings and change in attitude one would associate with transformation.[28] Familiarization, on the other hand, concerns making new ideas more comfortable.

Establishing Credibility by Laying Down Markers

In the automotive design context, an engineer's influence hinges in large part on a command of engineering language and ideas. This engineering language should be considered the "privileged" (Wertsch 1991) knowledge of the organization, given the greater influence it wields compared to other forms of talk. I describe the language used to demonstrate competence in privileged knowledge as "markers." Given the importance of engineering knowledge at AllCar, Dan's successful advocacy for the prototype team and other initiatives was in part a function of his ability to articulate knowledge so that it underscored his technical background. This was particularly important because his role in a coordinative engineering function put him at risk of not commanding the respect of his engineering colleagues. Dan's ability to lay down marks provided repeated and persuasive evidence of his engineering competence.

As G5 program manager, Dan was instrumental in orchestrating a significant and highly visible initiative to use the G5 as a pilot (referred to as a "learning line") with the objective of using the initiative as the basis for redesigning the automotive design process throughout AllCar. He was a strong advocate for the major innovations behind Virtual Car and single, simultaneous product and process release, which necessitated highly disciplined and coordinated division-wide completion of all of the G5's designs and manufacturing processes.

Specifically, the effort required design engineers to submit their part designs far earlier in the design cycle to establish design validity much sooner than in the past. Such early design validation was considered essential to faster cycle time and increased quality. This change in the design process required a substantial effort to initially persuade G5 managers and then frontline engineers to support the change on the front end. Just as important, it required post hoc information sharing and impression management to ensure both a positive assessment and continued organizational momentum. Part of that post hoc effort involved periodic meetings with managers from other divisions to review the initiative's results and reinforce the perception that the simultaneous product and process release had been a success.

At one such meeting, Dan met with a senior program manager (one rank above) from another division and the program manager's staff. Dan's sources warned him that the manager was skeptical about the success of the G5 experiments. On the basis of this information and his own observation of the manager from a distance, Dan indicated to me before the meeting that he expected the manager to find fault with the G5's simultaneous product and process release.

As it turned out, Dan correctly anticipated the manager's skepticism, but at the meeting he marshaled his considerable technical knowledge and verbal skill to explain what the G5 division had accomplished. As the excerpt displayed in the left column below indicates, he skillfully employed engineering jargon, laying down markers to advance the perception that G5's efforts had thus far been a success. The column on the right "decodes" the privileged knowledge that Dan deployed in his argument.

Quote	*Explanation of Terms*
SKEPTICAL SENIOR PROGRAM MANAGER: Some people were very impressed [with your efforts.] Let's separate fact from fiction. What's real?	
DAN: That's reasonable. [*Dan summarizes G5 accomplishments including "simultaneous interior/exterior BDA." At one point he indicates,*] We laid out all the DVPs and Rs, week 100 single and simultaneous design release.	The "simultaneous interior/exterior BDA" refers to a high-ranking meeting in which the integration of the car's internal and external design is reviewed and approved. "DVP and R" stands for "design, verification, plan, and report." The DVPRs are reports to track and coordinate the testing and verification of designs. "Week 100" (a fictitious number) refers to the number of weeks before the G5 manufacturing plant will be at full speed. An "engineering release" refers to a part design released only to the engineering community.
SKEPTICAL SENIOR PROGRAM MANAGER: How did you release? An engineering . . . release?	
DAN: Complete product release . . . APCN to CN. The old O1 cleanup. We moved O1 cleanup behind twenty weeks. But that's not the real	A "complete product release" refers to part designs that are completely final and submitted to manufacturing. "APCN," or Advanced

Quote	Explanation of Terms
key. We got the whole [expletive] car done. [The story is] how we got the whole car designed and done. The door line has commensurate door trim, windows, glass, and tracks, and seal moldings. . . . We flushed down these issues. F1 had been discovery. We said "bullshit". . . . I was a[n engineer] for twenty years . . .	Product Change Notice, is a form used to officially notify the engineering community of (and track) changes in engineering designs at the stage before they are released to manufacturing. "CN," or Change Notice, is a form used to officially notify the engineering community of (and track) changes in engineering designs at the stage after they have been released to manufacturing. "O1" is the design stage where parts are "production released," or final—a term from the old, and now outmoded, design process. "F1" is an earlier design stage where prototype parts are used to build test vehicles—also a term from the outmoded design process.

Dan's comments amounted to a powerful representation of G5's accomplishments. As impressive as the G5's new process had been, there were legitimate questions about how much had actually been accomplished. Dan's continuous stream of engineering terms laid down markers, establishing his engineering competence based on twenty years' experience, which he explicitly asserts at the end. By the end of the conversation, the skeptical manager relented somewhat, conceding, "I'm sure you did a better job [than your predecessors]."

Later that day, I pressed Dan about his defense of the simultaneous product and process release:

> ME: [The manager] was partly right, in the following sense. It is true, to a degree, that you are just incrementally improving the whole process. . . . It's like you're building a castle, and it's a beautiful vision on the hill, but the top three floors haven't been painted and don't have any furniture. But you emphasize the ideal of the castle to inspire people to keep moving toward what you believe is possible.
>
> DAN: That's right. But we are at a point that we cannot be ignored.

ME: Sure you could be ignored. You could have a lot of people like [the skeptical program manager] who discount what you are doing.

DAN: You know why we can't be ignored? We *have* built a castle. They had to come. [The senior executive] is telling his boss, "Look at these guys; you need to go see the castle." We've convinced him that a castle is being built.

The G5's simultaneous product and process release was a significant accomplishment that was tracked closely by the engineering community across the company. Dan's influential voice helped to generate initial support for the process innovation. His attention to guiding retrospective interpretation of these efforts both within and outside NewCar helped to solidify the status of simultaneous product and process release as an innovation and to maintain momentum for subsequent, related innovations.

The logic behind Dan's employment of "engineer talk," rich with markers that underscored his engineering experience and validated his assertions, was captured in his recounting of how he dealt with senior executive resistance when advocating for another process change:

> I sat down there on the edge of the table and said, "Damn it, Chris. You know, I am a product engineer." [The program management executive without direct engineering experience] wouldn't have been able to quell them. He wouldn't have been able to stop it. Well into it, [he] would have been stoned . . . and then the issue would have died.

> I was able to quell the stoning by saying, "Damn it, guys. I'm a product engineer. Been there, done that, managed product, managed people, been an engineer in management for well over twelve years. You guys knew me as a product manager over here in G4. I was a product manager on G3 and G2 in the early nineties. You can't tell me what you want me to do because I know. And blah, blah, blah, I'm not a [*expletive*] program manager. I'm a product engineer. You know?" That quelled them down and . . . the long and the short of it is, we worked through . . . a compromise between what I wanted and what it currently is, and we were able to get through that.

This exchange makes explicit Dan's motivation for presenting arguments in engineering language. In the previous conversation with the skeptical manager, Dan warranted his engineering knowledge more subtly. In this exchange, he made his engineering knowledge the center of his argument. The remarks about the consequences of not having engineering knowledge ("would have been stoned") underscored this awareness. Dan's flexible and adaptive use of language, including his laying down of markers, indicates his social skill in the context of knowledge articulation. Not all contexts exhibit privileged knowledge and the associated privileged languages (or jargon), but the culture at AllCar did. Skilled actors knew this and leveraged it whenever possible.

Linking Brokerage and Knowledge Articulation Processes

The types of knowledge articulation outlined above offer additional insights when viewed in relation to the three brokerage orientations. As noted, knowledge transfer is central to conduit brokerage (Table 2.2), which involves the movement of knowledge with little appreciable change from one node or location to another. In the simplest *tertius iungens* case, the *tertius* can make an introduction with little relational engagement. In the "brief *iungens*" (Obstfeld 2005) circumstance, in which an introduction is made with little or no accompanying relational work, no knowledge processes are implicated. In the case of Fred, Joann, and Gloria from Chapter 1, however, the cultivating of dyadic rapport (i.e., broker Fred connects first with Joann and then separately with Gloria) as a prelude to making a successful "triadic" combination with Joann and Gloria, would probably involve translation, and quite possibly transformation at multiple points.

Table 2.2.   Making Knowledge Relevant across Brokerage Orientations

	Conduit	*Tertius gaudens*	*Tertius iungens*
Knowledge transfer: Movement of knowledge, largely unchanged, between people or social worlds	**Knowledge transfer inherent in conduit brokerage**	**Blocking or filtering knowledge transfer**	No inherent knowledge process in "brief *iungens*" introductions, but transfer possible with sustained *iungens*
Knowledge translation: Alternative framing of knowledge to bridge different understandings between people or social worlds	Possible but not essential	**Filtering and shaping of knowledge to constrain or alter knowledge flow, or to cultivate conflict**	**Translation quite likely to cultivate the potential for collaboration between two alters or expand the collaboration to new parties with heterogeneous preferences**
Knowledge transformation: Significant resynthesis or new synthesis of shared knowledge and practice	Possible but unlikely	Filtering and shaping of knowledge to constrain or alter knowledge flow, or to cultivate conflict	**Transformation quite likely to cultivate the potential for collaboration between two alters or expand the collaboration to new parties with heterogeneous preferences**

SOURCE: Original table.

Tertius gaudens brokerage exhibits a more complex set of potential knowledge processes. In one scenario, the *tertius gaudens* blocks or restricts the exchange of knowledge in order to keep alters apart to protect an idea or opportunity, suggesting the absence of knowledge articulation. Alternatively, the *gaudens* may filter information (Burt 1992) for the same purpose, to keep alters apart, foster competition between alters, or regulate the level of competition or partnership in order to limit or control the opportunity that might emerge. This filtering suggests a more active regulation of knowledge movement along the back stage to front stage or complex to simple dimensions (or even a move to make simple knowledge more complex). Skillful brokerage should be understood as emerging out of a blend of all three brokerage orientations; it necessarily involves different mixes of transfer, translation, and transformation.

Where do we locate knowledge transfer, translation, and transformation in the above examples? Rick's more straightforward requests (e.g., "Could this be thicker?" and "I was hoping this would be lighter") are syntactic and thus involve transfer, in the sense that there is little or no ambiguity about their meaning.[29] At NewCar, the engineers' rendering of knowledge often involved translation in order to make it relevant to the socially embedded understandings and interests of their interlocutors. Translation was in evidence when Rick, out of necessity, added metaphors (e.g., "Barbie doll stipple") and analogies (e.g., "tight like a scuba outfit") to clarify meaning and necessary design activity (i.e., action) that would bring the next prototype closer to realizing his design objectives. As noted earlier, this perspective taking that underpins discourse is also a function of social skill. Rick's exchange with the engineering group also illustrates how a given communication can involve blends of knowledge transfer and translation.

Disagreements stem from a localized chunk of knowledge and interest that is at odds with another localized chunk of knowledge or interest. Brad has knowledge and interests, originating with the expertise and practice situated within the Body division, that make his resistance to Paper Car sensible. Dan has his own knowledge and interests—originating with his expertise and practice of program management and his responsibility for the design and production of the entire G5 automobile—that cause him to advocate for the fullest implementation of the comprehensive (albeit clunky) Paper Car database. Dan's reframing of NewCar as an "information company" (rather than a car company) is an effort to transform shared knowledge by reshaping Brad's perspective on his design work in order to align it with his own.

Conclusion

The emphasis in this chapter on knowledge articulation opens up new opportunities for theorizing about the knowledge processes necessary for mobilizing action to get new things done. The fleeting, extemporaneous nature of knowledge articulation and the associated liminal status of much organizational knowledge, located somewhere between fully tacit and highly codified forms, together suggest that the articulation of knowledge is an emergent process, for which codification is far less frequent than prior research might suggest. More often, articulation episodes involve the surfacing of tacit understanding that leaves no lasting trace. Such knowledge may be spoken but not written down, may be written on whiteboards that are erased without record, or may be sketched on pads that are subsequently tossed in the trash.[30] Informal representational activity is essential to innovation but cannot be evaluated in terms of codification or breadth of diffusion. Much like the brokerage process, knowledge articulation, while influential, is as difficult to observe as it is fleeting, and often without a physical trace.

Many prior treatments of knowledge process privilege codification and diffusion. Past work on knowledge transfer (e.g., Kogut and Zander 1992; Zander and Kogut 1995), for example, emphasize codification and diffusion as natural endpoints of articulation processes. More recent approaches suggest that influential artifacts or codifications, such as boundary objects (Carlile 2002) or formalized heuristics (Tsoukas and Vladimirou 2001), are key to problem resolution or are the engine for knowledge creation. For these approaches to be fully realized, they must be coupled with more temporary knowledge articulation processes that may never "evolve" into anything resembling formal representations or heuristics but critically facilitate the movement and application of knowledge.

I hope to shift attention from *representations* to *representing* and the means by which representing influences action. Such representing implicates social skill, by which mutual intelligibility, persuasion, and enlistment are essential to develop and mobilize support for new innovation. Knowledge articulation involves varying levels of intent and effectiveness. Actors with varying levels of social skill direct their articulation efforts to influence the conduct of others in their local contexts.

Effectiveness in influencing action requires making knowledge more explicit, useful, and relevant to pressing problems faced by a specific audience and within a specific context. Suchman reflects this perspective when she observes,

"The efficacy of plans, instructions, and the like . . . relies precisely upon the ability of those who make use of them to find the relation of . . . general prescriptions to [a] particular occasion that faces us now" (2003, 301). Usefulness and relevance make knowledge articulation persuasive, linking it to action, innovation, and the coordination necessary to innovate. These properties describe the conditions under which articulated knowledge is most likely to influence others' disposition to act and, in turn, to promote enlistment in innovation projects.

The ongoing enlistment and coordinative action necessary for innovation is arguably the single most important articulation outcome in organizing to get new things done. As Giroux and Taylor suggest, knowledge justification is part of "a complex process, which involves the enrollment of interests to form coalitions of sufficient strength to inflect the organizational intention in the direction of support for . . . innovation" (2002, 498).

This treatment of knowledge articulation—as ubiquitous, fleeting, combinatorial, and intertwined with organizing—suggests a need to use with care the knowledge-creation terminology that has become central to knowledge-based theorizing. The emphasis here is on incipient knowledge-articulation episodes, involving the bracketing, surfacing, framing, and reblending of meaning (Gavetti 2005). These more subtle acts involve translation and transformation by moving between back stage and front stage; between complex and simple; among past, present, and future; by balancing familiarizing and defamiliarizing; and by laying down markers. Such an approach clarifies the importance of the actor or broker to knowledge creation and movement.

Knowledge articulation involves managing knowledge through communicative and coordinative skill rather than through disembodied ideas involving knowledge capture, storage, and retrieval (Carlile and Rebentisch 2003; Spender 2005). Such a perspective shifts focus from the identification, diffusion, and protection of knowledge assets to the social skill needed to promote and coordinate small- and large-scale innovation efforts. As Winter indicates, "The failure to articulate what is articulable may be a more severe handicap for the transfer of knowledge than tacitness itself" (1987, 172). This alternative perspective also suggests the merits of cultivating and rewarding managers capable of skillfully practicing knowledge articulation and creating the innovation outcomes (routine- and project-based) that such activity makes possible.

3

Creative Projects

In the Introduction, I argued that the origins of getting new things done, whether related to innovation, strategic firm adaptation, new venture formation, or collective action, involve variations on a "small numbers" game, which requires mobilizing people and resources around imagined future states. Having examined brokerage in Chapter 1, and knowledge articulation in Chapter 2, I now turn to the BKAP model's dependent variable to which these activities are directed: *organizational innovation*. Organizational innovation can arise from one of two trajectories: routine-based innovation and project-based innovation.

Like individuals, organizations are largely, but not exclusively, creatures of habit. They follow routines. Organizational routines involve the maintenance, extension, or adaptation of a repetitive, relatively stable pattern of shared action. Feldman and Pentland define organizational routines as "repetitive, recognizable patterns of interdependent actions, carried out by multiple actors" (2003, 95). Routines account for the vast majority of organizing (Nelson and Winter 1982) in the same way that habit characterizes a vast percentage of individual action (Wood, Quinn, and Kashy 2002). The brokerage and knowledge articulation behaviors explored in earlier chapters function to support the iterations inherent in routines' established patterns of action.

Yet while such routines often serve to perpetuate existing organizational activity, they can also be engines of organizational innovation when they generate new, or improve existing, products and services (e.g., new product development, total quality management) (Winter 1994). Innovation-generating routines constitute the dominant form of routine-based innovation. A secondary form of routine-based innovation arises from routine-modifying organizational change. Here the existing routines themselves are changed but the routines' objectives or goals remain largely intact. In both these cases of routine-based innovation, the routine's inherently repetitive nature constrains the expression of innovation.

Project-based innovation (or *creative projects*), by contrast, involves nonroutine innovation that arises from either transforming existing routines or launching substantially new action. The latter effectively starts with a blank slate (e.g., greenfield corporate projects, new entrepreneurial ventures). In the former, the innovative action seeks to disrupt, substantially remake, or replace an existing routine instead of adapting it. A thought-provoking example is suggested by Winter's imaginary popsicle factory:

> The president says to the director of research that "I've just been reading about the great strides that have been made recently in cryogenic technology. Why don't you look into that and see if there is anything that will give us a cheaper way of making popsicles." This directive . . . set[s] in motion a chain of organizational events which may ultimately lead to a new set of production processes. It differs [from more incremental adjustments to routines] in that the timing, magnitude, and character of its impact on the way the firm operates is very hard to predict. (2006, 130)

In this case, the president launches a course of action, or trajectory (Strauss 1993), to achieve his projected outcome: a never-before-seen, cryogenic-based popsicle production process. This trajectory represents a form of change markedly distinct from the stability or iterative change associated with routines. Creative projects, as in this example, involve the attempt to introduce change through an evolving vision or projection of a new end state, and the pursuit of that projected end state through emergent action. The novel action trajectory the president launches, with its multiple actors and contingencies, contrasts with the incremental innovations that are more closely associated with routines.

The organizational literature, despite extensive references to nonroutine work, still struggles for an explanatory construct that addresses the president's action, the projective alternative to the iterative organizational routine. The "creative project" fills that gap, offering a construct that accounts for an important source of organizational change and change in routines.

Creative projects offer an alternative trajectory to routine-based innovation for getting new things done. I assert that this trajectory is disproportionately responsible for getting new things done, and highlights the importance of nonroutine action within an expanded view of innovation and organizing. Research on the nonroutine is sorely underdeveloped in social science, yet the potential for providing an account of innovative action that takes place outside the routine is invaluable. I define creative projects specifically as *emergent trajectories of interdependent action, initiated and orchestrated by multiple actors to introduce change into a social context* (Obstfeld 2012).[1] Note the deliberate resemblance between my definition of the creative project and Feldman and Pentland's definition (2003) of the organizational routine: both are trajectories, or patterns, of interdependent actions carried out by multiple actors. Both involve knowledge and networks. The creative project, however, is defined not by repetition but by *emergent* action to introduce change.

The creative project is not an exotic unknowable; in fact, routines and projects are closely related. A creative project may stabilize into a routine as a newly emergent path of action becomes more repetitive over time, and a routine may evolve into a creative project, whether in response to an unforeseen disruption or a perceived failure in performance. Nevertheless, the routines construct cannot adequately address the ways significantly new things get started in organizations, no matter how much it is stretched or redefined. As Emirbayer and Mische point out, "Human actors do not merely repeat past routines; they are also the inventors of new possibilities for thought and action" (1998, 983–84).

I now distinguish, with more precision, between project-based and routine-based innovation trajectories. First, the creative project's newly envisioned end state features an outcome that constitutes a distinct break from previous action as compared to the organizational routine's end state, which bears a strong resemblance to the previous repetitive outcomes of the routine. Second, while both trajectories comprise a sequence of tasks, creative project subtasks are linked to a wider range of possible subsequent subtasks. Put differently, creative projects involve more choice both with respect to sequencing and among a broader range of possible tasks as compared with organizational routines, which have a well-established set of tasks linked in more predictable ways. Third, participation in creative projects involves a greater number of choices regarding a broader range of potential participants than the more defined participation characteristic of organizational routines. Notice how this increased participation necessitates more active brokerage. Finally, the combination of greater task and participant uncertainty in a creative project necessitates more deliberation about the choices for how to link tasks and participants than found in organizational routines. Notice how this increased deliberation necessitates more active knowledge articulation.

Theorizing this underexamined, nonroutine trajectory for getting new things done is the focus of this chapter. First, I will briefly draw on insights from pragmatist philosophy with respect to the interplay of routine and nonroutine action. Next, I will summarize the organizational literature's treatment of routine and nonroutine innovative action and their expression in the learning-curve construct. I next introduce a conceptual framework for action trajectories in project-based and routine-based innovation. I then explore, in terms of that framework, the role of brokerage and knowledge articulation in creative projects. This is followed by a brief examination of the implications of this framework on meta-routines, where I introduce the idea of meta-trajectories. Next, I provide exploratory criteria for making more concrete distinctions between innovation in creative projects and innovation in organizational routines. Finally I present a hypothetical "Apple Watch" case that illustrates many of the concepts introduced here.

A Pragmatist View of Nonroutine Action

Determining where new action comes from poses a daunting challenge for social scientists and philosophers of action alike. One answer emerges from a long line of work in the pragmatist tradition, which takes as its point of departure the breaking away from habits or routine activity, a break that Joas refers to as "situated creativity":

> Action constantly encounters unexpected obstacles: goals show themselves to be unattainable; simultaneously pursued goals prove to be mutually exclusive; attainable goals have doubts cast upon them by other actors. In these various crises of habitual action, the action situations have to be *redefined* in a new and different way. . . . Hypotheses are put forward: suppositions about new ways of creating bridges between the impulses to action and the given circumstances of the situation. Not all such bridges are viable. . . . For the pragmatists, action consists not in the pursuit of clear-cut goals or in the application of norms, and creativity is not the overcoming of obstacles along these prescribed routes. Anchoring creativity in action allows the pragmatists to conceive of creativity as precisely as the liberation of the capacity for new actions. (1996, 133)

The above scenario offers an account of what Joas refers to as "new actions." They emerge when habitual action encounters unexpected obstacles. These obstacles produce crises that stimulate impulses to resolve those crises, ultimately leading to new actions. In other words, actors are driven to new actions in part because preexisting goals and the usual means for achieving them don't address the crisis.

The routine/nonroutine interplay to which Joas alludes has deep intellectual roots. Herbert Simon (1945), citing Stene (1940), described organizational routines as an analogue to individual habit, which serves to conserve "mental effort by withdrawing from the area of conscious thought those aspects of the situation that are repetitive" (1945, 88). Reflecting on habit, Simon acknowledged the work of pragmatists John Dewey and William James, whose perspectives allowed for both the importance of habitual action and the significance of repeated interruptions of habitual action. Dewey, the pragmatist philosopher and social theorist, explored this break from habit quite explicitly in a passage that invites comparison to Joas, above:

> Life is interruptions and recoveries. . . . A novel factor in the surroundings releases some impulse which tends to initiate a different and incompatible activity. . . . In this period of redistribution impulse determines the direction of the movement. It furnishes the focus about which reorganization swirls. Our attention in short is always directed forward to bring to notice something which is imminent but which as yet escapes us. Impulse defines the peering, the search, the inquiry. . . . As organized habits are definitely deployed and focused, the confused situation takes on form, it is "cleared up"—the essential function of intelligence. . . . With conflict of habits and release of impulse there is conscious search. (cited in Adler and Obstfeld 2007, 23)

From such a pragmatist perspective, action is "anchored in unreflected belief in self-evident facts and successful habits" (Joas 1996, 129) but is reconstructed in response to changing aspects of the world—a reconstruction that itself will eventually take root as a new, nonreflexive habit. Strauss, acknowledging this pragmatist tradition, argued that "there is a process whereby routine plays into creativity and innovation, which in time flows back into the realm of the routine" (1993, 206). This sequential interplay between habit and its emergent interruptions is noted in the sociological and organizational literatures (Louis and Sutton 1991; Mead 1938; Snow et al. 1998; Tilly 1997).

Although the central role of routines is well developed and theorized within organization theory, nonroutine action, while recognized at critical junctures, has received shorter shrift. Given its critical importance to getting new things done, nonroutine action demands new attention, in parallel with inquiries into organizational routines, at several levels of analysis.

Having contrasted the two fundamental action trajectories—organizational routine and the creative project—I now want to introduce three underlying facets of action, to further our understanding of how these action trajectories vary and how they unfold in pursuit of innovation. Emirbayer and Mische (1998) argue that all agency can be decomposed into three temporal elements: (1) iterational or routine (past oriented), (2) practical-evaluative or improvisational (present oriented), and (3) projective (future oriented). Stable organizational routines can be located on the repetitive end of a continuum, oriented toward the past; creative projects occur on the opposite end, oriented toward the future; and a range of adaptive routines are situated in between (Figure 3.1). All social action, according to Emirbayer and Mische, contains a blend of these three temporal elements within a chordal triad, where one "note" is sounded more forcefully than the others. A routine may well have elements of improvisation and projection, and a projective trajectory, in which a set of actors initiates some new collaboration, can certainly include routine and improvisational elements.

Over a half century before Emirbayer and Mische, the pragmatist philosopher John Dewey offered a similar tripartite framework for individual action, which he described as consisting of habit, intelligence, and impulse (Dewey 2002; Adler and Obstfeld 2007; Winter 2013), a framework that Adler and Obstfeld (2007) tied to organizational correlates (Figure 3.2).[2] In Dewey's scheme,

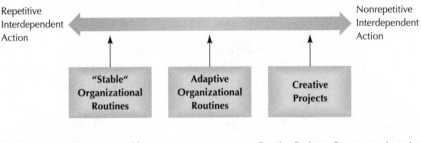

Routines–*Repetitive, recognizable trajectories of interdependent actions, carried out by multiple actors.* (adapted from Feldman and Pentland 2003)

Creative Projects–*Emergent trajectories of interdependent actions initiated and orchestrated by multiple actors to introduce change into a social context.* (Obstfeld 2012)

Figure 3.1.   Create Projects versus Routines on a Repetitiveness Continuum
SOURCE: Original figure.

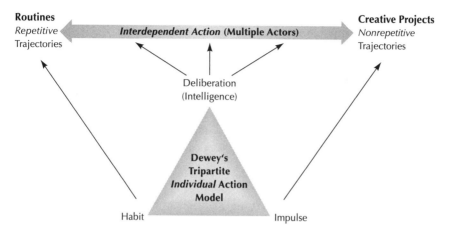

Figure 3.2.   Locating Project-Routine Trajectories in Dewey's Tripartite Action Model
SOURCE: Original figure.

habit, like Emirbayer and Mische's iterative dimension, corresponds closely with organizational routines; and impulse, like Emirbayer and Mische's projective dimension, corresponds quite closely to creative projects (Adler and Obstfeld 2007). The projective dimension represents an impetus for change and deviance that may redirect routines or alter their content. The third dimension in Dewey's scheme, intelligence, corresponds closely with Emirbayer and Mische's practical-evaluative or improvisational dimension. It can be understood to correspond with deliberation, playing a central role in priority setting and direction setting, as well as management decision making (Adler and Obstfeld 2007; Winter 2013).

Deliberation, although not constituting a separate action trajectory, plays a critical role in shaping both creative projects and routines. According to Dewey, it is a form of thought experiment:

> Deliberation is a dramatic rehearsal (in imagination) of various competing possible lines of action. . . . Deliberation is an experiment in finding out what the various lines of possible action are really like. It is an experiment in making various combinations of selected elements of habits and impulses, to see what the resultant action would be like if it were entered upon. But the trial is in imagination, not in overt fact. The experiment is carried on by tentative rehearsals in thought which do not affect physical facts outside the body. (2002, 123)

Deliberation, in short, is a reflection in action, which may be individual or shared and is critical to the unfolding of innovative action in both routines and projects.

In exploring the interplay of Dewey's three dimensions on the individual level, Winter (2013) gives the example of driving on "automatic pilot" as pure habit; of responding to challenging driving conditions by choosing a new route as a blend of habit and impulse; and of paying heightened attention to detail and cues in order to navigate that unfamiliar route as a blend of habit and deliberation. With these three components of action in mind, let's return to the contrast between the organizational routine and the nonroutine to explore what organizational theory has emphasized and what it has overlooked.

Distinguishing between the Routine and the Nonroutine
Summary of the Organizational Literature

The organizational routines literature has staked out a well-established role in organizational theorizing (Simon 1945; March and Simon 1958; Cyert and March 1963; Nelson and Winter 1982). The central role of repetitiveness in routines (Becker 2004; Cohen and Bacdayan 1994; Gersick and Hackman 1990; Pentland and Rueter 1994) is reflected in Feldman and Pentland's (2003) definition of routines as "repetitive, recognizable patterns of interdependent actions, carried out by multiple actors." In addition to the repetitive nature of routines (Nelson and Winter 1982; Feldman and Pentland 2003), there is general agreement that routines are accomplished by multiple actors via some form of cooperative relationship (Cohen and Bacdayan 1994; Feldman and Pentland 2003); have defined ends and means (Pentland 2004) insofar as the existence of a routine implies relatively predictable inputs, intermediate steps, and outputs (Gersick and Hackman 1990); and involve more tacit (Nelson and Winter 1982; Polanyi 1958), unarticulated, or procedural knowing (Cohen and Bacdayan 1994).

A "stability" perspective is well developed in the foundational work on organizational routines (Simon 1945; March and Simon 1958; Cyert and March 1963), and an extensive line of subsequent work has explored the difficulty of changing routines (Edmondson, Bohmer, and Pisano 2001; Gersick and Hackman 1990; Gilbert 2005; Levitt and March 1988; Nelson and Winter 1982; Orlikowski 2000). Even though this earlier literature noted that routines were adaptable and modifiable, only recently has the "adaptive" capacity of routines received extensive attention (Feldman 2000; Feldman and Pentland 2003; Pentland and Rueter 1994; Weick, Sutcliffe, and Obstfeld 1999). This more recent conceptualization emphasizes actors' ability to consciously and incrementally alter routines from within over time. Feldman (2000), for example, found such plasticity and adaptability in university dormitory hiring and training routines, as have others (Miner and Estler 1985; Miner 1991; Howard-Grenville 2005; Cardinal, Sitkin, and Long 2004) in other contexts. Feldman and Pentland's (2003) influential reconceptualization of routines suggests that they are a source not only of stability and inertia but also of flexibility and considerable change originating from within the routine itself.

While not as central to organizational theorizing, the nonroutine has been addressed by several preeminent organizational theorists in recent decades. The organizational literature has long recognized that some things get done through non- or less-routine action, where repetition is not a guide on what to do next. Early on, for example, Simon (1945) argued that organizing needed to be understood in terms of routines and decision making (Cohen 2007), and March and Simon (1958) spoke of unprogrammed tasks, while others have addressed unstructured decision processes (Mintzberg, Raisinghani, and Theoret 1976), adhocracy (Mintzberg 1993), nonlinear work systems (Pava 1986), nonroutine processes (Lillrank 2003), routines and sensemaking (Weick 1995), and stretch goals (Sitkin et al. 2011).

Organizational work has also explored forms of coordinated action closely related to, but distinct from, routines. Burns and Stalker, for example, postulated a continuum of mechanistic to organic forms of organizing, with the former emphasizing clear hierarchy and specialist roles suitable for stable conditions and the latter being responsive to fluid conditions with "fresh problems and unforeseen requirements for action" (1961, 121). The move along the continuum toward more organic organizing is characterized by an increase in lateral communication and action, in which job specialization gives way to work with "less formal definition of methods, duties, and powers" (6). This organic form involves a "network structure of control, authority, and communication" and a "preparedness to combine with others" (121, 125). The authors observed that managers' adaptation to these less-defined organic forms of work often elicited an anxiety that lifted them out of complacency and rigidity, a complacency often associated with routines (e.g., Gersick and Hackman 1990).

Also, from a firm-level perspective, Perrow (1967) analyzed organizational technology in terms of the number of exceptional cases that the technology presents and the type of analysis undertaken when such exceptions occur. Nonroutine technology involved more exceptions and a less logical, less systematic form of analysis to address them. The analysis associated with nonroutine technology, according to Perrow, draws upon "the residue of unanalyzed experience or intuition, and relies upon chance or guesswork" (1967, 196). The individual search associated with nonroutine work (which Perrow explicitly associates with Burns and Stalker's organic form of work) involves experimentation and "feel," coordination by feedback (versus plan), and diminished distinctions between supervisors and staff. Perrow's nonroutine work resembles the mutual adjustment and selective decentralization found in Mintzberg's adhocracy (1993).

Nonroutine action also loosely corresponds to the exploration aspect of the exploration/exploitation distinction described by March, in which exploration involves "search, variation, risk taking, experimentation, play, flexibility, discovery, innovation," and exploitation involves "refinement, choice, production, efficiency, selection, implementation, [and] execution" (1991, 71). The relationship of nonroutine action to exploration and exploitation is complex. Farjoun (2010), for example, makes a key distinction between stability and change in mechanisms and outcomes. In important cases, he points out, *change-based* mechanisms can detect and *stabilize* variation, as in the case of high-reliability organizations (Weick et al. 1999), and *stable* routines can support *change*, as in the case of meta-routines (Adler, Goldoftas, and Levine 1999).

Alongside exploration and exploitation, a second conceptually distinct organizing dimension, according to March (1994, personal communication), is the extent to which organizational action is rule driven, at one extreme, or alternatively driven by anticipation, calculation, and rule making, at the other. According to this perspective, exploration could be rule driven, as in the case of the meta-routine, or display nonroutine anticipation and calculation. It is in this latter form of exploration that the creative project would fall.

Sensemaking also speaks to the tension between routine and nonroutine as captured in Weick, Sutcliffe, and Obstfeld's observation that "sensemaking begins with chaos"—that is, unfamiliar events and tasks that don't lend themselves to easy interpretation (2005, 411). In the face of such breakdowns that involve the departure from routines, individuals exchange provisional understandings to restore a "consensually constructed, coordinated system of action" (Taylor and Van Every 2000, 275). Note the similarity between this account and the one posed by pragmatist philosophers noted earlier in this chapter. Part of that process of consensually constructing interpretation and action, according to Weick et al. (2005), involves the noticing, bracketing, and labeling of experience, and the dialogic articulation, elaboration, and enlistment in a given interpretation and the ensuing action that such interpretation suggests. While much recent empirical study of sensemaking is situated within routines, sensemaking as a social phenomenon is ultimately most pronounced around the disruption of ongoing routinized action.

Scholars have explored how innovation involves the coexistence of routine and less-routine forms of action (Brown and Eisenhardt 1997; Burgelman 1991, 1994; Jelinek and Schoonhoven 1993; Miner, Bassof, and Moorman 2001).

Newer research explores temporary, project-based organizations such as film production, engineering, and construction firms (Baker and Faulkner 1991; Bechky 2006; Gann and Salter 2000; Grabher 2002; Hobday 2000; Lundin and Söderholm 1995; Prencipe and Tell 2001; Turner and Keegan 1999, 2001), but this research generally makes such distinctions at the firm level, where the characteristic microsocial action can vary widely between the routine and nonroutine. This line of literature on the less- and non-routine is important because it makes clear that routines are a necessary but not sufficient construct for establishing a complete picture of organizing. The many accounts of nonroutine action are also conceptually unparsimonious and in some sense ad hoc, being added onto rather than integrated with the foundational routines literature.

Insights from the Learning Curve

Most routines, however well established, emerge during a "preroutine" period, when new action is first conceived and pursued. Due to its comparatively fleeting duration, that preroutine period is frequently treated as a footnote to the routine it spawns rather than as a crucial, fertile source of new action worthy of greater theoretical and empirical scrutiny. To contrast the "initial learning" associated with new nonroutine action to the learning that takes place within routines, let's turn to a time-honored concept: the learning curve.

The core insight yielded by the learning curve, that a unit cost of production typically decreases as experience accumulates and decreases at a decreasing rate (Argote 2012), is primarily an insight about the evolution of organizational routines. Here I won't focus on the relatively predictable progression of within-routine efficiency over the course of a learning curve; instead, I want to focus on the distinct period of experimentation before and after the first execution of a given routine in a learning-curve progression, which is a period of nonroutine-to-routine transition. The "before" concerns all the messy interactions that led up to the routine—and to a lesser degree the period between the first and the second iterations of the routine, which comprises a uniquely intensive consolidation of coordination and anticipation, and within which the most learning within the routine typically takes place.

Consider, for example, a learning-curve progression involving the falling cost of airplane production—the first widely used application of the learning-curve concept within organizations. Costs fall as the organization uncovers successively more efficient ways to coordinate airplane production within established routines. But imagine the activity that preceded the production of the first unit of a given airplane coming off the line in a newly built factory. If we go far enough back in time, we would almost certainly discover a crisis or disruption, perhaps when the then-current model failed to match the competition's airplanes on price or performance, or had become technologically outmoded. This disruption might have spurred a vision of a new airplane model with updated design and manufacturing processes. There would have been considerable deliberation, probably some of it sustained and passionate, regarding how the new model's design and manufacture would be pursued. In addition, while such a preroutine period would have involved a great deal of data collection and analysis, one can imagine how the "whether" and "how" of the new production line would initially have been far more subject to deliberation and guesswork—in other words, the contemplation of various possible lines of action that Dewey describes.[3]

In the preroutine period, an important accompanying decision process might have concerned whether and how to go about building an entirely new manufacturing facility with advanced technology and a just-in-time parts supply. We would naturally bracket and exclude such manufacturing design activity from our learning-curve production time calculations, but of course all of these messier pre- and early-production period issues would have carried enormous significance.

Such an early stage constitutes a local or firm-level analogue to what, at an industry level, Tushman and Anderson (1986) refer to as a period of discontinuous change. This involves a dramatic break from existing practice, followed by an "era of ferment," when fundamentally new alternatives compete to replace outmoded technologies (see also Rosenkopf and Tushman 1994). The first plane assembled, with its associated novelty, would likely account for both the greatest inefficiency as compared to subsequent iterations, and the most dramatic learning and cycle-time improvement as a result. The first and the second executions of the production routine are also those iterations that least resemble the routine's subsequent repetitions. During this early period, those operating the routine will likely exhibit a newly emerging cognitive orientation that reflects a transition from *configuring* to *operating* the routine, and eventually to operating the routine more efficiently.[4]

On the back end of the learning curve, when executives reach the conclusion that certain advances in technology demand the development of a new plane, we would expect an eventual departure from the efficient routine altogether and, after considerable deliberation, the initiation of a new assembly process and production curve. That shift would involve a new and different projected outcome (e.g., the design and manufacture of a more technologically advanced airplane), a break from more-routine action, a move to a new production arrangement, and after another relatively nonroutine period, the reestablishment of a production function that would presumably again display routine-based learning.

We may argue about the point at which to start the learning-curve calculation and how to account for the jump from one learning curve to another, or when dramatic leaps in technology and corresponding changes in coordination transpire, but the intersection of the routine and the nonroutine is an inherent and vital part of organizing. Understanding the tension between the routine and the nonroutine is central to understanding how innovations and organizations get started, survive, and scale.

Conceptual Framework for Action Trajectories

This review of the organizational literature on routines and the nonroutine is thus far relatively broad and nonspecific with respect to action. To address that deficit, I turn to explore in greater depth the concept of the creative project, introduced earlier in the chapter. Next, I will place it, along with the routine, underneath the conceptual umbrella of the *trajectory*—a sequence of interdependent actions involving multiple actors (Strauss 1993). And finally, I will elaborate how brokerage and knowledge articulation help to explain mobilization of innovation within creative projects.

Conceptual Background

Creative projects involve the attempt to introduce change initially identified through a vision of a new end state, followed by the pursuit of that projected end state through emergent action. The creative project construct applies both to a substantially new path of action within an organization and to an early-stage entrepreneurial start-up, in which the entrepreneur embarks upon a relatively uncharted course of action to achieve a newly conceived projected outcome, a new organization. Phenomenological and sociological theory identifies the project (or *entwurf*) as a unique reflexive act establishing "in-order-to," or future-oriented, motives (as opposed to "because," or past-oriented, motives) to act (Schutz 1967).[5]

The routine's hallmarks of repetitiveness and recognizability (Feldman and Pentland 2003) can be adapted to further distinguish between organizational routines and creative projects. The central questions of organizing (and sensemaking) posed by Weick et al. (2005)—"What's going on here?" and "What do I do next?"—align with these recognizability and repetitiveness hallmarks, respectively. Organizational routines and creative projects can be distinguished by the extent to which the answers to these questions are known as action unfolds. An organizational routine is defined by its repetitiveness, where the answer to the question "What comes next?" is relatively well in hand. For the creative project, on the other hand, because the action is novel, the answer to the question "What comes next?" is not as clear. The creative project's lack of repetitiveness fosters less recognizability and suggests that the associated activity, participation, is also less well defined. As a result, the answer to the recognizability question, "What's going on here?" is not as clearly defined. Conversely, the repetitiveness that defines a routine fosters a recognizability that allows for its labeling and for more defined participation, which in effect supplies a more concrete answer to the question "What's going on here?"

Creative projects are conducted de novo, or "as if for the first time" (Merriam-Webster). This is consistent with Amabile's (1996) characterization of heuristic as opposed to algorithmic tasks. Amabile indicates that while for algorithmic tasks "the path to the solution is clear and straightforward," for heuristic tasks there is no "clear and readily identifiable path to solution" (35). The objective, and the means by which that objective is pursued (although not whether it is ultimately realized), identify a project as "creative." This approach views routine and nonroutine action in the form of the organizational routine and the creative project as a difference in degree, not in kind (Adler and Obstfeld 2007). Although innovation in organizational routines and that in creative projects are functionally different, where the former serves to ensure continuity while the latter supports markedly distinct change, I argue that they can be understood within the same conceptual and theoretical framework. This approach enriches the organizing literature by bridging the conceptual gap between routine and nonroutine action, making it possible to examine and compare action across an innovation continuum ranging from the routine to the nonroutine.[6]

Trajectory Strategy: Projection and Scheme

As noted at the beginning of this section, by "trajectory" I mean a sequence of interdependent actions involving multiple actors, based on Strauss (1993, 53). As early as 1926, Schumpeter saw routines as paths of action or trajectories (Becker, Knudsen, and March 2006).[7] I employ the trajectory concept to connect the literatures on the routine and nonroutine around a concrete unit of analysis. Such an integrated perspective, I argue, presents new opportunities to clarify the mechanics of nonroutine action and to consider the sources of and distinctions between different forms of organizational stability, change, and innovation.

Strauss et al. allowed for trajectories to range from "quite routine to highly problematic," seeing the trajectory as involving "a course of action, but also embrac[ing] the interaction of multiple actors and contingencies that may be unanticipated and not entirely manageable" (1985, 11). This range of flexibility essentially mirrors the continuum of stable routine to creative project and repetitive to nonrepetitive action. Since the trajectory concept accommodates both the more stable and more adaptive forms of routines as well as the creative project's emergent interdependent action, it serves as a unit of analysis for observing and comparing a range of action responsible for stability and less-routine forms of innovation and entrepreneurship. Although I emphasize the distinction between the innovation in projects and routines, it is important to remember Dewey's insight that deliberation figures centrally in the unfolding of both project and routine trajectories.

In addition to his basic definition of the trajectory, Strauss (1988, 1993) provides a useful taxonomy of trajectory-related concepts (Table 3.1). On the basis of Strauss (1993), I define a *trajectory projection* as a vision of an expected or desired outcome. The trajectory projection is the central element that motivates and guides a trajectory; both the creative project and the routine are motivated by the trajectory projection, though with different implications. Given the organizational routine's reliance on past iterations, the projection, while malleable,

Table 3.1. Conceptual Framework for Action Trajectories: Innovation through Creative Projects and Organizational Routines

	Creative Projects Action pursuing a newly envisioned outcome	**Organizational Routines** Action following an established routine
Trajectory Strategy		
1. ***Trajectory Projection:*** vision of an expected or desired outcome	***Envisioned*** based on a new possible future	***Established*** based on past repetition
2. ***Trajectory Scheme:*** plan designed to guide interactions in pursuit of the trajectory projection	Conscious, ***explicit***	More ***tacit***
Trajectory Management		
1. ***Brokerage Activity:*** action connecting and coordinating people, organizational units, and resources	Unfolding coordination through **novel combinations** of people, organizational units, and resources	Ongoing coordination of **preestablished combinations** of people, organizational units, and resources
2. ***Knowledge Articulation:*** action making knowledge explicit, usable, or relevant to the situation at hand	**More explicit** pursuing and adapting trajectory projection and scheme in response to the situation at hand	**More implicit** increasing articulation to address disturbances (or contingencies) to get work back on track

Source: Obstfeld 2012, 1574.

is comparatively well formed and defined. For the creative project, on the other hand, the trajectory projection has greater importance in initiating action and far more susceptibility to being reshaped as trajectory action unfolds.

For the creative project, the trajectory projection implicates a novel vision or goal, the pursuit of which disembeds action from prior routines and sets it in motion.[8] A crisis may well have motivated that projection, but the projection itself is what motivates and guides future action. Unlike the kind of fixed, externally determined goal that characterizes the routine, a creative project's trajectory projection evolves as actors encounter the opportunities, challenges, and surprises associated with its pursuit. Dewey (1916) referred to this activity as "ends-in-view," involving a malleable target or end state that guides activity as that activity unfolds.

From this perspective, the trajectory projection is still relatively undefined (or at least not completely resolved) and only becomes more specified as choices regarding certain means are made—again, choices that are likely made continually as the project trajectory unfolds. Principal among these continual choices are those made with respect to the actors and ideas to be orchestrated through the brokerage and articulation mechanisms explored in previous chapters, and to which I will return shortly. I refer to this coevolution of the goal and the means for approaching that goal as "evolving means and ends."[9]

Given the tentativeness of the creative project's envisioned end state and the actors' approach toward that end state, Strauss provides other constructs, including the *trajectory scheme*, a plan to realize that vision. I view the trajectory projection and trajectory scheme in concert as defining the trajectory strategy. Strauss also introduced *trajectory management*, the action employed to direct the trajectory through all of its phases by deploying the evolving trajectory scheme. Trajectory management depends on the effective use of brokerage and knowledge articulation discussed in previous chapters, which I address further below.

These concepts can be used to think about trajectories with more precision and to further distinguish between routines and creative projects. To refer once again to Dewey's tripartite action model, this is where Strauss in effect allows for the deliberation and improvisation present in organizational routines but more pronounced in creative projects. This language also allows us to distinguish between the trajectory projection that motivates and guides a creative project and the nonroutine action that unfolds *in pursuit of* that trajectory projection (which quite possibly will reshape the trajectory projection as action unfolds). The creative project's nonroutine action consists in large part of reflection about different combinations of people, ideas, and resources by which to pursue the trajectory projection, and how those combinations could best be accomplished. Keep in mind, however, that the vast majority of organizational action consists of organizational routines that target relatively stable, predictable trajectory projections established in previous iterations of those routines and involving preestablished combinations of people, ideas, organizational units, and resources.

From the standpoint of Dewey's tripartite action framework, deliberation and improvisation are centrally involved in both routines and creative project trajectories, but in a qualitatively different way. The distinction to be made between different levels of deliberation can be made by referencing Weick's (1998) examination of improvisation, which he defines as "reworking precomposed material and designs in relation to unanticipated ideas conceived, shaped, and transformed under the special conditions of performance, thereby adding unique features to every creation" (Berliner 1994, 241). Notice how such reworking "in performance" relates to both the adaptive routines described by Feldman and Pentland (2003) as well as the creative project's mobilizing of existing networks and knowledge in pursuit of new projections.

The distinctions Weick makes between different degrees of improvisation offer clues to the kind of deliberation characteristic in routines versus projects. Weick describes four levels of improvisation, starting with "interpretation" (i.e., when people take minor liberties with a melody) and "embellishment" (where the melody is rephrased but recognizable), and contrasting them with "variation" (involving the insertion of clusters of notes not in the original melody) and "improvisation" ("involving transforming the melody into patterns bearing little or no resemblance to the original model") (1998, 545, after Berliner 1994).

In a similar fashion, deliberation in the case of organizational routines involves comparatively smaller improvisational liberties; otherwise, the trajectory would no longer feature the repetition characteristic of a routine. Alternatively, the deliberation found in creative projects involves more significant departures from previous action, that is, new forms of action that target new means and ends.

The trajectory projection itself supplies the impetus for this disembedding by providing new answers to the organizing questions, "What's going on here?" and "What do I do next?" and in so doing, provides a lever for new action. The extent of the break from past action (i.e., novelty) is in part a function of the ambitiousness or even the "foolishness" (March 1971) of the projection, as well as subsequent circumstances beyond the actor's control that may constrain or propel a set of actions forward toward new outcomes.[10] The trajectory projection introduces a new aspiration that motivates and guides new action, as opposed to targeting a previously established trajectory projection associated with a routine.

Trajectory Management: Brokerage and Knowledge Articulation

The project/routine framework, at this point, needs to be linked to the brokerage and knowledge processes discussed earlier. From the BKAP model's perspective, brokerage and knowledge articulation account for coordinated or mobilized action necessary to routine-based or project-based innovation (creative projects). On the basis of my ethnographic work (Obstfeld 2012), I characterize an action trajectory's interdependent action, whether in the form of an organizational routine or a creative project, as consisting of brokerage activity and knowledge articulation. This brokerage and knowledge articulation activity is employed by actors developing trajectory schemes which anticipate action and, perhaps most importantly, for trajectory management, which engage action. When associated with the creative project, such trajectory management, like the full-on improvisation described by Weick, "is much tougher, much more time consuming, and places higher demands on resources" (1998, 545). In the case of the creative project, coordination unfolds through more novel combinations of people, drawing on one or more of the three brokerage orientations described in Chapter 1, but especially the *tertius iungens*. Within the organizational routine, ongoing coordination also involves multiple brokerage orientations but is more likely to involve preestablished combinations of people forged in previous iterations of the routine, and thus combinations that are less novel and effortful.

Correspondingly, knowledge articulation as described in Chapter 2 is central to both creative project and routine trajectories but serves a more active role in drawing new combinations of people (as well as ideas and resources) together in the creative project by providing a common purpose, either in terms of the ultimate trajectory outcome or in terms of important intermediate steps necessary to propel the project forward. In a routine, where the trajectory is more repetitive and thus predictable, knowledge articulation helps execute or justify (versus persuade or enlist on behalf of) an iteration on a preestablished combination or outcome; as such, it is often more implicit or tacit, with the exception of situations where work is off track.

Where the creative project is concerned, managing contingencies (or disturbances) in the pursuit of a novel trajectory projection is commonplace and defines the trajectory as actors shape, anticipate, or respond to the inherently unpredictable unfolding of action. In routine-based innovation, the path carved out through prior repetition constitutes the ground against which the occasional disturbance or interruption (or contingency, in Strauss's lexicon) is managed. Contingency management, in the case of the routine, concerns the efforts associated with getting or keeping work on track.

From Meta-Routines to Meta-Trajectories

Thus far we have kept the distinction between the organizational routine and the creative project relatively simple, matching the creative project as a unit of analysis to a basic level of analysis associated with the seminal organizational routines literature. I refer to this basic level of analysis as a first-order trajectory, consistent with Cohen and Bacdayan's (1994, 555) definition of routines as "patterned sequences of learned behavior involving multiple actors who are linked by relations of communication and/or authority"—that is, involving handoffs between a relatively finite, tractable number of actors. Such first-order trajectories involve (loosely) between two and two dozen actors.

A meta-routine has been described as a routine for changing routines (Feldman and Pentland 2003), or a standardized problem-solving procedure to improve existing routines or create new ones (Adler et al. 1999). One classic example of such a meta-routine is found in total quality management practices that promote a systematic investigation of preexisting frontline routines (e.g., Plan-Do-Check-Act or PDCA) for the purposes of decreasing their inefficiencies (Winter 1994). The meta-routine construct (Cyert and March 1963; Nelson and Winter 1982) is premised on a hierarchical configuration of routines that concerns how certain higher-order routines guide change in subordinate operational routines.

Meta-routines may also refer to routines of a broader scope that contain constituent, subordinate routines. In this latter sense, simple or first-order routines involving the interactions among individuals often cluster and intersect in larger, overarching meta-routines. The automotive design process, for example, constitutes a meta-routine consisting of a dozen or more simultaneously executed, subordinate design routines. Such a design meta-routine ultimately leads to a production meta-routine (also comprising dozens of first-order routines) in which automobiles are actually manufactured.

Meta-routines therefore guide or connect more than one constituent or first-order routine. In light of our theorizing about projects and routines, however, meta-routines may be better conceptualized as a variant of a broader category of meta-trajectories. With this adjustment, we expand our language for describing how meta-trajectories coordinate and integrate action within and across organizations. The meta-routine that directs first-order routines is by now well established. A meta-project might consist overall of a novel trajectory projection and nonroutine action, but quite possibly of a mix of subordinate first-order routines and projects. The idea of a creative project built or stacked on routines was anticipated by Strauss who indicated, "Even in the most revolutionary of actions the repertoire of routines does not vanish" (1993, 195). Consider, by way of illustration, a band of revolutionaries who as part of the process of fomenting the revolution (a meta-project), gather daily at a coffee shop (first-order routine) for their morning meeting.

At the meta level, a creative project might have first-order routines as its constituent elements; alternatively, a meta-routine might not only produce change in subordinate routines, but also foster the generation of creative projects. I will return to such considerations in a subsequent chapter with respect to collective action and dynamic capability. For our purposes here, let's consider creative projects as either first-order trajectories or, alternatively, as second-order trajectories whose signature is nonroutine pursuit of a trajectory projection, but which may have one or more subordinate routines as part of their functioning. The mix of first-order routines and creative projects in such a meta-trajectory may evolve over time, changing the project/routine nature of the meta-trajectory itself.

Distinguishing between Innovation in Projects and Routines: Explanatory Criteria

Having described the broad theoretical distinction between innovation in organizational routines and creative projects, I would now like to explore in greater depth a precise operationalization of the distinction between these two innovation trajectories briefly noted earlier in this chapter. To make this distinction, I stress the combination of people, ideas, and choices. I refer to these, with the addition of resources (Baker and Nelson 2005) and artifacts, as *elements*.[11] By *resources*, I have in mind things such as money, time, rank, and status. By *artifacts*, I have in mind physical or digital objects that help organize interactions and interpretations. (The "master cross-section" in the next chapter's case is just such an artifact.)

I emphasize strategic actors' active linking of elements in pursuit of their goals (i.e., trajectory projections) over time. In this view, organizational routines and creative projects involve sequences of events, or combinations, that strategic actors orchestrate through the linking of people and other elements over time. A strategic actor links people with other elements (e.g., a problem, a solution, and two people he or she judges to be interested in the problem/solution combination). This suggests that a person can be both a strategic actor and an element in another strategic actor's trajectory activity. Note how knowledge articulation would govern the skill with which the strategic actor introduces the problem and solution so that it generates interest from the people involved.

Network theory is helpful in demarcating the relationship of elements (e.g., in open or closed, or large or small networks) and the process for combining them (e.g., brokerage processes, especially *tertius iungens* linking). In an organizational routine, the universe of possible elements and the sequence in which they are combined are, by definition, constrained. There is an established set of participants and a sequence by which they are joined. In more open creative project trajectories, by contrast, a broader universe of potential elements is linked at each event in the trajectory sequence.

The criteria provided below further delineate the distinction between routine-based innovation and project-based innovation. These criteria assume that (1) both organizational routines and creative projects involve coordinated action between multiple actors; (2) this coordinated action entails trajectories comprising a series of tasks linked together over time (Malone et al. 1999); (3) each task involves choices regarding which tasks to execute and which elements to combine; and (4) the elements involved include multiple actors, problems, solutions, resources, and artifacts.

The assumption here is that both organizational routines and creative projects are constructed from elements and discrete actions, or "tasks," that connect those elements in various configurations. Basic task forms include transformation, tie or relationship formation, and brokerage or combination. A transformation involves action that takes an input and transforms it to some output. Such a transformation may alter an artifact, a relationship, a collaborative unit, a resource, or all of these. By tie formation, I mean the creation of a relationship between two people as a prelude to combination; and by combination, I refer to the formation of temporary or more permanent clusters of three or more people, through the *iungens* and *gaudens* orientations described in Chapter 1, along with other elements such as ideas and resources. (For my purposes, I will restrict "combination" to the linking of three (rather than two) or any number of more elements.)

How might such dynamics unfold over time? A five-person cluster may meet in the conference room to explore a proposal, one task involving combination. This is followed by another task: a larger meeting with the board and an investor (a second combination). Two team members may have met prior to the five-person meeting (a prior task). Any of these episodes could constitute a transformation.

Regarding elements, we take different actors who engage in different combinations as a point of departure. As noted, combinations and other tasks, in addition to involving people, may also involve other elements, such as problems (e.g., underperforming processes), solutions (e.g., a new technology that might remedy the underperforming process), resources (e.g., money and time), and artifacts.

Both organizational routines and creative projects involve sequences of tasks linked together over time. If we draw from "garbage can" theory (Cohen, March, and Olsen 1972), trajectories involve choices as to which tasks to execute and which elements to join. Within an organizational routine, calling a meeting (e.g., a department meeting) might constitute a choice of a task in which the participation is somewhat prescribed. Within a creative project, calling a meeting (e.g., to interest a set of potential stakeholders) might involve considerable deliberation regarding whom to invite, when to invite them, and what to discuss (and to avoid discussing).[12]

With this model of tasks and elements in mind, I now provide criteria meant to capture with more precision the distinctions between organizational routines and creative projects. Consider four criteria that distinguish between innovation through organizational routines and creative projects (Table 3.2). Criterion 1 concerns the trajectory projection, the goal of trajectory action. For the organizational routine, the trajectory projection concerns a familiar endpoint, which bears a strong resemblance to previous outcomes of the functioning routine. For the creative project, the trajectory projection concerns an outcome unrelated to previous action. Trajectory projections may vary according to the extent they differ from prior action.

Criterion 2 concerns the linkage among tasks within a trajectory. Creative project tasks (versus organizational routines) are linked to a wider range of possible future tasks. In its simplest form, each subsequent organizational routine task is well specified, whereas for a creative project, the tasks are not predetermined and are likely to vary among a range of possibilities. Creative project tasks have a greater potential to alter subsequent determination of tasks as well as the end goal, and less predictable causality among subtasks. In other words, subtask A is less likely to yield a certain result or lead to subtask B in creative projects than in organizational routines.

Criterion 3 holds that creative project tasks involve consideration of a greater possible universe of elements from which agents involved must select.[13]

Table 3.2. Explanatory Criteria: Distinguishing between Innovation through Creative Projects and Organizational Routines

Explanatory Criteria	Distinguishing between Creative Projects and Organizational Routines
1. Trajectory Projection	The creative project's trajectory projection features an outcome unrelated to previous action as compared to the organizational routine's trajectory projection, which is based on past repetition.
2. Subtask Linkages	Creative project (versus organizational routines) subtasks are linked to a *wider range of possible future subtasks.* In its simplest form, each subsequent organizational routine subtask is well specified; for a creative project, the subtasks are not predetermined and are likely to vary among a range of possibilities. Creative project subtasks exhibit: • *Greater potential to impact determination of other prior, concurrent, or subsequent subtasks* as well as the end goal. • *Less predictable causality among one another.* In other words, subtask A is less likely to yield a certain result or lead to another subtask B in creative projects than in organizational routines.
3. Trajectory Elements	For creative projects, there exists a greater *universe of potential elements.* • Creative project subtasks involve *more uncertainty as to the elements* that might be selected. (Organizational routine elements are more likely to be known or preestablished.) • Creative project subtasks combine, on average, *a greater number of elements.* • Creative projects involve a *greater level of interaction among elements.*
4. Deliberation	Creative projects involve *more active deliberation* about the nature of each subsequent subtask and which elements to join, as well as more conscious combinatorial work to link, articulate, and coordinate elements.

SOURCE: Original table.

In a given creative project, for example, an agent will consider a greater range of people necessary to secure support in order to move the project forward. In the case of the project, support is not predetermined and the nature of the support necessary to move the project is unclear. As a result, there is a greater level of experimentation around tie and relationship formation. There are three features to criterion 3. First, creative project tasks involve more uncertainty as to the elements that might be selected. (Organizational routine elements are more likely to be known or preestablished.) Second, creative project tasks combine, on average, a greater number of elements. This is, again, a function of the greater level of experimentation and trial and error necessary to move a given creative project forward, as compared to an organizational routine. Third, there is also a greater level of interaction among elements. This is in part a function of (1) the greater number of elements in play for a creative project, (2) the search for combinations that move the project forward, and (3) the potential to bring new elements into play as the trajectory projection evolves.

The increased number of alternatives and the increased uncertainty (addressed by criteria 1 through 3) give rise to increased deliberation and active consideration suggested in criterion 4. Criterion 4 proposes that creative projects involve more active deliberation about the nature of each subsequent task and which elements are to be joined, as well as more conscious combinatorial work to link, articulate, and coordinate elements. This criterion is consistent with a central insight that routines economize on conscious thought and deliberation (Simon 1945; Cyert and March 1963). This suggests an extension of a theory of routines that includes creative projects, while remaining faithful to the original Carnegie School's reason for why routines exist: to economize on cognitive effort. Again, note the correspondence here to deliberation in Dewey's tripartite scheme (habit, deliberation, and impulse). Here this suggests that a combination of impulse and deliberation predominate in a creative project trajectory. This list of criteria suggests the distinctions, as well as the close relationship, between creative projects and organizational routines. Future work could refine, apply, and test these criteria, as well as identify new criteria.

The criteria presented here suggest both a precision and a fluidity between creative projects and routines. The first and most obvious relationship between creative projects and routines is that of evolution between creative projects and organizational routines, considered in a previous section. With the above criteria in mind, we can speak with more precision about the idea of an initially unscripted creative project, with many alternative choices of actors, ideas (problems and solutions), and artifacts, evolving into an organizational routine as the breadth of choice at each juncture narrows. I described such a relationship earlier in exploring learning-curve dynamics, suggesting that creative projects often evolve into stable organizational routines. With learning-curve dynamics in mind, we can also point to the opposite relationship, in which a well-established organizational routine often eventually gives way to a new project, which takes as its objective the "improvement" (or supplanting) of an existing routine.

Examination of the role of choice in constructing trajectories suggests fundamental differences between the search associated with an organizational routine and a creative project. Specifically, in an early definition of an organizational routine, March and Simon indicate:

> We will regard a set of activities as routinized, then, to the degree that choice
> has been simplified by the development of a fixed response to defined stimuli. If
> search has been eliminated, but a choice remains in the form of clearly defined
> and systematic computing routine, we will say that the activities are routinized.
> (1958, 163)

In this view, removing the search for what to do next in response to a given situation defines a routine. The criteria posited here illustrate the more constrained search associated with organizational routines. From this perspective, the "search" associated with problem solving is not only about the sorting of alternatives; it may also involve an ongoing deliberation about how to combine various people, problems, solutions, choices, and resources, punctuated by the provisional linking of these elements in new combinations in order to move creative project trajectories forward. It also underscores the importance of knowledge articulation as a means for linking problems to solutions, binding people to ideas (either problems or solutions), and enlisting people to participate in unfolding action trajectories.

Apple Watch Case

The longitudinal nature of the organizational routine is not hard to imagine, as in the case of Apple's design routine for creating a next-generation iPhone, or the corresponding manufacturing routine. When Apple creates a next-generation iPhone, there exists a relatively defined, somewhat predictable sequence of activities to follow as well as a clearly defined group of actors to involve. The trajectory projection is to transform updated iPhone design elements (a larger, brighter screen, a faster processor, waterproofing, and a more powerful camera) into a next-generation iPhone design, and subsequently to transform updated iPhone components into next-generation iPhones at an Apple manufacturing facility. The next-generation iPhone represents a relatively limited change to an existing trajectory projection and associated routines.

The longitudinal construction of a creative project trajectory, however, while somewhat intuitive, involves an unscripted trajectory assembled from a sequence of combinations that merits at least some description. Imagine the circumstances surrounding the emergence of an altogether new Apple product, such as the Apple Watch.[14] In an interview, Jonathan Ive, the senior vice president for design at Apple responsible for the Apple Watch, described a period when the Apple Watch became a formal project, though one that was "still tentative and very fragile" and unique: "It's not very often that we start something that's an entirely new platform" (Parker 2015, 134). This was no doubt a period full of deliberation. While one might expect some of the participants (Ive, Steve Jobs, and CEO Tim Cook), the participation of many others was presumably far more tentative as Apple navigated the uncertain vision for the Apple Watch's design, software platform, and manufacturing. In contrast to the iPhone upgrade, a wider range of internal Apple technology and design experts presumably was needed to determine new technologies and aesthetics central to design and product launch success.

With many new actors "in play," proposed meetings often involved new combinations of novel and established actors. Presumably many meetings were contemplated and abandoned, while others were held. On the basis of deliberations at meetings held, certain potential collaborators were ruled out, perhaps due to past history, reputation, or lack of fit, while others were ruled in. Among the dozens of such exploratory meetings, Apple might have called on outside designers involved in upper-end watch design, advanced materials, emoji art, and novel technologies useful to Apple for the first time. Some meetings advanced the project, while others led only to dead ends. Maybe one outside collaborator was not sufficiently up to date on newly developing technology, and another seemed insufficiently creative, while a third, who appeared to be an excellent fit, couldn't make herself available for the requisite engagement.

A fourth outside designer might only be able to join the project if his colleagues were included as well. This designer's unique contribution in helping Apple achieve the Apple Watch's not-yet-fully-formed objectives might hold such promise that Apple would decide to hire the designer's entire external design team as well. Such a decision, in turn, would lead to new, unforeseen combinations of actors and ideas brokered by senior Apple product managers.

My fictitious (and necessarily primitive) discussion of the Apple Watch's design process attempts to depict the nature of decisions arising initially in the project, as well as the basis for making those decisions. I can only imagine that there were multiple trajectory paths, some of which involved combinations that went no further and others that led to sequences of *other* combinations, and paths that led to new features or manufacturing choices.

It is also likely that a trajectory projection, perhaps the vision of a technology-infused and aesthetically arresting watch, guided and motivated the initial action. Even after the viability of the Apple Watch trajectory projection was established, it no doubt substantively evolved but still continued to motivate and guide action. If Apple weren't as secretive as it is known to be, an even broader range of internal and external actors might have been invited to participate in the emerging design effort. Finally, the abundance of resources at Apple's disposal influenced the creative project trajectory, as compared to the form it might have taken if it had been a start-up in the hands of two Stanford undergraduates. In the latter case, for example, more time would have been required for meetings with investors. Furthermore, as a result of operating outside of a well-established corporate context, the Stanford students might have engaged a far broader range of actors and opinions. Finally, keep in mind that Apple likely will eventually release an Apple Watch 2 and 3, at which point its design and production will be comparatively predictable and exhibit all the hallmarks of an organizational routine.

My main objective in this example has been to characterize the trajectory as a sequence of fluid, brokerage-based combinations of people and resources, as well as an evolving articulation of ideas—projections—by actors as they seek to mobilize organizational support around the development and introduction of a product, the Apple Watch. Certainly, the broker plays an enormously influential role, or succession of roles, in this fluid sequence.

Conclusion

This introduction to the creative project begins to address the largely unresolved tension in organization theory concerning how organizations get new things done amid the relative stability of everyday operations, by putting forth a new conceptual framework that integrates routine and nonroutine action. The framework presented here, involving routine and creative project trajectories consisting of different forms of trajectory projection—combinatorial action linking people, problems, solutions, choices, and resources; knowledge articulation; and contingency management—can be used to understand how organizational routines and creative projects differ, interrelate, and in some cases evolve into one another.

As noted, social theorizing has disproportionately emphasized routines, in part because they make up the lion's share of what we do individually and collectively, and because their inherent repetitiveness increases the probability that they will be observed, retained in individual and shared memory (Cohen and Bacdayan 1994), theorized, counted, and represented as variables in statistical models of organizing. Nonroutine activity often receives less attention from practitioners and academics due to its fleeting nature and the challenge posed for systematic observation (Emirbayer and Mische 1998). Projects are more susceptible to going unobserved, or rapidly and imperceptibly fading from view, making them more difficult to capture and analyze. The creative project illuminates new possibilities for nonroutine coordinated action with implications at the micro, firm, and even field levels.

In sum, the creative project is introduced as an alternative trajectory form—a form of interdependent action conceptually distinct from but closely related to both stable and more innovation-oriented cases of organizational routines. Creative projects are exploratory ventures that offer one means by which organizations and their routines change. By addressing the relationship between these two constructs, this theoretical framework expands our understanding of how organizations change and innovate, while identifying the role of agents in directing more- or less-routine action trajectories. Change and innovation may occur through existing routines or by setting a nonroutine action trajectory in motion. These nonroutine trajectories consist of substantially new projections of alternative future outcomes and undetermined, emerging combinations of people and other elements engaged in the trajectory action to achieve those outcomes.

Previous scholarship has recognized, and sometimes endeavored to reconcile, the coexistence of more- and less-routine forms of organizing in the form of paired constructs of organic and mechanistic organizing (Burns and Stalker 1961), or exploration and exploitation (March 1991). Alternatively, scholarship has employed umbrella constructs that reconcile similar dimensions at a higher level, constructs such as semistructures (Brown and Eisenhardt 1997), quasi-formal structures (Jelinek and Schoonhoven 1993), or ambidexterity (Tushman and O'Reilly 1996; Gibson and Birkinshaw 2004). Others (Perrow 1967; Strauss 1988) have presented unifying frameworks that connect routine and nonroutine action, but to date the literature has lacked an empirically based, action-oriented framework that addresses the rising importance of nonroutine work at a microsocial level.

By differentiating action within the firm using the trajectory framework, newly emerging and ongoing forms of innovation and change are illuminated. Creative projects, for example, may be causally linked to but remain distinct from the routines they seek to change or spawn. Creative projects display a unique capacity to disrupt the activity that came before as well as to scale. Most creative projects do not result in field transformation, but nevertheless display, at the microlevel, the emergent mobilization driven by social skill that gets new things started.

The introduction of the creative project relieves the strain created by attempts to understand a broad range of organizational activity in terms of routines alone. Such efforts stretch the concept of routine so broadly that the boundaries of the construct have become difficult to define, according to some observers (Cohen et al. 1996; Foss 2003; Weick 1998; Weick and Sutcliffe 2006).

Creative projects are not a new form of organizing. In fact, project-based action has always constituted a source of organizational adaptation and competitiveness. Yet such organizing takes on new importance in a "flattened" world. In the Introduction, I referenced Boltanski and Chiapello's emphasis on a new way of organizing, involving a firm "featuring an organisation that is very flexible; organised by projects; works in a network; features few hierarchical levels; where a logic of transversal flows has replaced a more hierarchical one, etc." (2005, 165). Constellations of creative projects that actors engage in a project-based world also suggest an underlying form of learning without strict repetition that is increasingly central to new forms of distributed, project-based work. The learning found in such creative projects is consistent with the formulation of new objectives and action strategies found in double-loop learning, as opposed to the incremental adjustments to ongoing action found in single-loop learning (Argyris and Schön 1978).

The construct of the creative project anchors a disparate literature on non-routine action, while fitting that literature within a broader framework that explicitly ties nonroutine to more-routine forms of organizing. The inductively derived conceptual framework (based on data and analysis given in Chapter 5) presented here is as relevant to routines in general and innovation-oriented organizational routines as it is to the markedly distinct change found in creative projects. Taken together, the framework and findings extend our current understanding of how organizations get new things done, and how creative projects and routines differentially unfold through unique forms of trajectory projection and schemes as well as through interdependent action comprising brokerage activity, knowledge articulation, and contingency management.

The BKAP framework into which both routine-based innovation and project-based innovation fit has applicability at multiple levels within and across organizations. It can provide a fuller theoretical account for the unfolding of action trajectories involving changes in processes, the creation of entirely new products or services, and the founding of firms. I provide several applications in Chapter 7 to illustrate the breadth of arenas to which this model may contribute insight.

4

Mobilizing for Routine-based Innovation: NewCar's Manual Shifter Redesign Initiative

Having laid out a theoretical framework and initial evidence for the role of social network brokerage (Chapter 1) and knowledge articulation (Chapter 2) in creative project- and routine-based innovation (Chapter 3), I now turn to an extended ethnographic case and related survey data to show how network and knowledge processes interact to produce innovative action over time. In this chapter, I draw upon these data from NewCar to argue that network and knowledge dynamics are central to routine-based innovation. Field observations focus on how three socially skilled actors orchestrated knowledge and networks to mobilize action for innovation within the automotive design routine.[1]

I first provide relevant context for the automotive design process, after which I walk through the extended case in three phases of activity and analysis. The first phase involved disruption of the existing design routine and the initial challenges experienced by the manual shifter "Crunch Team" in its efforts to respond to that disruption.[2] This phase was characterized by confusion and a relative absence of knowledge sharing and coordinative action. I then present a second phase which contrasts two engineering efforts to mobilize support for innovation, one that failed (Eric's "Whalebone") and the other that succeeded (Henry's "Double Boot"). In the third phase, I also provide a contrast between two efforts to mobilize support; however, this time the advocacy came instead from designers (Sam who failed, and Alex and Joe who succeeded) for a design solution, the "master cross-section." Alex and Joe used their novel design approach to reconcile multiple stakeholder interests and knit together several major design changes. In both phase 2 and phase 3, successful innovation advocates mobilized action through brokerage and knowledge articulation to get new things done.

Context: The Automotive Design Process at NewCar

As a rule, automotive design is not a particularly inventive process. While key actors exhibit substantial creativity in their mobilization of networks and knowledge, they ultimately act within a prescribed set of routines with clear boundaries, objectives, and time horizons. Outside of the earliest design stages, innovative action generally involves the introduction of focused, incremental changes or the overcoming of smaller disruptions to keep or get the routine back on track. This changes substantially, however, when a design issue threatens the successful completion of the overall design routine.

Since NewCar's part-design engineers were responsible for creating automotive parts that ultimately led to the parent company's (AllCar) all-important automotive design efforts, these engineers held substantial power and status. For AllCar, then, as for the rest of the American automotive industry, design engineers were at the top of the automotive status hierarchy. They were organized into five engineering divisions that corresponded with core automotive functions: Body, Chassis, Electric, Interior, and Powertrain. The other engineering departments, such as Vehicle Development and Program Management, which coordinated design activities, along with Marketing and Manufacturing, had less influence.[3] The status associated with part design was reflected by one design engineer's blunt critique of another nonengineer on a project: "Not an engineer, never released a part, never built a car."

Engineering design work at NewCar relied heavily on a myriad of cross-functional teams of design engineers with responsibility for a given part or subassembly of the automobile. Within the G5 organization, these cross-functional teams continually reviewed the compatibility of intersecting designs originating out of different functional areas. Within these teams, each design engineer was typically paired with a designer, who worked under the engineer's supervision to prepare continually updated digital designs reflecting the latest engineering changes. Cross-functional teams consisted of roughly a half dozen or more engineer/designer combinations, whose designs interfaced along different parameters, and included representatives from such nondesign areas as Manufacturing and Marketing.

A new automobile's general design parameters were determined relatively early in the automotive design process. Subsequent design issues often revolved around what engineers referred to as "packaging," critical, ongoing mutual engineering changes in part design to ensure that the parts fit and functioned together within a fixed, often extraordinarily small, three-dimensional space. Packaging issues were resolved through an iterative process that usually lasted for many months. Packaging negotiations were an expected, and potentially contentious, part of the design process, because they often required engineers to compromise on designs they had spent months, if not years, perfecting. It was widely understood that engineers became more resistant to changes later in the design process, as accumulated time invested in and commitment to a series of subsequent, related design decisions increased.

Phase 1: Design Disruption and Initial Challenges
Creation of the Manual Shifter "Crunch Team"

Twice a month, approximately fifty of NewCar's senior managers and engineers gathered at the Executive Design Review Meeting (EDRM).[4] Their goal was to confront a handful of critical engineering issues regarding the design of a completely new automobile, the G5,[5] on which specific engineers and their managers were asked to report. Except for senior managers, whose seats at the table were always available or were quickly made so, those who arrived late to the EDRM generally wound up standing in the back. Presenters stood before the senior and middle managers seated around a thirty-five-foot conference

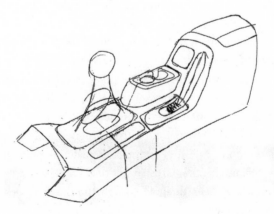

Figure 4.1. NewCar Designer's Early Visualization of Console with Manual Shifter
SOURCE: Unofficial figure from defunct automotive firm, 1988.

table, with other engineers and managers filling the seats along the walls of the crowded room. EDRM attendance often exceeded fifty people. According to one engineer, "This meeting is for high shooters. You usually come in here when you are in deep shit, and tap dance."

In fact, many of the presentations at the EDRM that day went quite well. The measure of a successful presentation was how well the presenters persuaded the senior managers that the issue in question was well understood, well managed, and on track to be resolved. A presentation's effectiveness could be judged by the number of questions it generated—fewer was better—and the ease with which the questions were answered. Successful presentations were well organized, brief, and convincing, often benefiting from a display of data or a coherent narrative account signaling that the problem was under control.

The manual shifter/console design presentation, one of many scheduled for this particular EDRM, concerned the use of a relatively new simulation model, the "move macro," which had been developed to anticipate the impact that various driving scenarios (e.g., making a hard right turn) would have on different design areas of the G5. In this case, the move macro was being used to anticipate driving scenarios that might cause the manual shifter to bump into the surrounding console. (Figures 4.1 and 4.2 provide an early designer's sketch of the manual shifter and console, and a clay model at a later stage of development, respectively.) When the simulation was linked to NewCar's state-of-the-art computer-aided design (CAD) software, which displayed parts or subassemblies constructed from geometrically accurate components in colorful 3-D space, a sequence of shifter "moves" showed how the shifter would be displaced under

Figure 4.2. Later-stage Clay Model of the G5 Manual Shifter
SOURCE: Unofficial figure from defunct automotive firm, 1988.

thirty-eight different driving conditions. This particular EDRM presentation was meant to show, first, that the simulation predicted a number of problem moves, and second, how they could be addressed. The problem moves were easy to spot in the display shown to the group, as the virtual shifter appeared to crash through the virtual console walls that surrounded it. Such instances were termed "interferences" or "hits" (see Figure 4.3 for an illustration of an interference in the Virtual Car software).[6]

The two managers making the EDRM presentation, Clay and Seth, represented the two main departments involved in the shifter/console design effort: the Powertrain group, responsible for the design of the automatic and manual shifters; and the Interior group, responsible for the design of the plastic molded "box" or console that surrounded the shifter. Because the engine movements did not affect the inherently simpler design of the automatic shifter, the two presenters hoped to persuade the EDRM to separate the console design efforts for the automatic and manual shifters. This would allow the design work for the automatic transmission to go forward while the group focused its problem-solving efforts on the interferences associated with the manual shifter. The

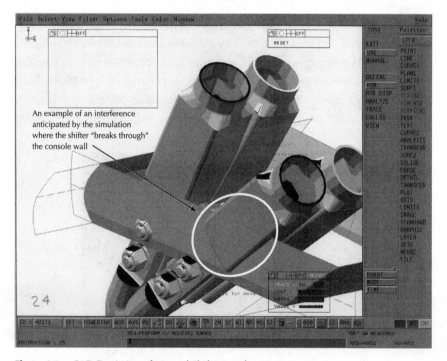

Figure 4.3. CAD Depiction of Manual Shifter Interference

Source: Unofficial figure from defunct automotive firm, 1988.

decision to separate the designs of the console and proceed along separate tracks would cost several hundred thousand dollars for additional tooling that would be needed to manufacture a second console.

The knowledge articulation task for Clay and Seth was to communicate their initial analysis, so that its results were clear, useful, and relevant to the situation at hand, to the EDRM audience tasked with defining the next steps for the manual shifter design. The simulation being used to evaluate the shifter movements generated 38 moves for each of the G5's planned four engine options—or 152 possible moves. Of the possible moves, only a small subset created interferences. Clay and Seth's decision to show all 152 moves tried the patience of the executives in the room, who shifted uncomfortably in their seats, and it obviously violated the "complex-to-simple" dimension of knowledge articulation. The presenters took approximately twenty-five minutes to complete their presentation, easily twice the length of a typical EDRM presentation. Simply stated, the two presenters did not succeed in this effort.

Several comments signaled a growing skepticism and impatience in the EDRM audience, and by extension the impending failure of the knowledge articulation effort. At one point, for example, an observer expressed his confusion: "It's difficult to determine which [driving condition] leads to which engine movement." In response to another question, Clay offered to rerun a portion of the simulated shifter moves, at which point a senior manager, signaling his dwindling patience, proclaimed, "Oh my God, no!" to considerable laughter. An executive engineer said, "My impression is that [we've provided] the biggest [physical] opening for a manual transmission in the history of the world. . . . Something is wrong here. Something doesn't compute." Further questions led to a facetious remark by NewCar's executive program manager: "You could just mount [the manual shifter] on the roof," another comment that was met with derisive laughter. At the end of the presentation, the program manager asked, "Would it solve all the problems?" Seth responded, "I wouldn't say that."

The program manager abruptly ended the discussion, rejected the split console request, and announced that they were moving on to the next scheduled presentation, indicating, "I do not think the solution is to . . . go to a new console." He tasked Clay and Seth with continuing the work of their newly formed "Crunch Team," a term for task forces convened to resolve high-priority design issues. They were directed to find a single console design solution or a more compelling argument for splitting the consoles when they next came before the EDRM. After the meeting, the program manager commented on the presenters' lack of credibility: "It's like you having gone home and your wife calling you 'Fred' [when your name isn't Fred]. It's disconcerting. . . . It speaks to a fundamental disconnect."

This scene at the EDRM illustrates a number of complex knowledge articulation challenges. The success of Clay and Seth's presentation hinged on their ability to generate credibility both for their account of the problem and for their recommended solutions. Their imperative was to create an understanding of the issue at hand (mutual intelligibility), acceptance for the feasibility and attractiveness of the opportunity posed by their proposed solution (persuasion), and support for the proposed design approach (enlistment). Instead, the presentation generated confusion and doubt about the presenters' credibility and competence. Several weeks later, the G5 program manager still remembered with frustration the failed manual shifter presentation, dismissing it as a "soft soap, 30,000 feet overview."

My attendance at the manual shifter EDRM presentation just described alerted me to the new crunch team's challenges, at which point I began attending many of this cross-functional team's meetings. These meetings served as a basis for the present case on the relatively late-stage disruption of the G5 design process, which required a collective effort to innovate in order to solve a pressing design problem against a backdrop of a fairly well-structured organizational routine.

Crunch Team "Slippage" and Lack of Mutual Intelligibility

At the time that the move-macro simulation identified the alleged problem with the G5's manual shifter design, a majority of the engineering design work had already been completed, under the direction of the Interior division (responsible for the console) with close participation from Powertrain (responsible for the shifter). As noted earlier, there were a number of problematic "interferences," in which the virtual shifter would crash through its surrounding console.

The crunch team created to address these interferences was a loosely organized group of approximately thirty engineers, designers, and manufacturing representatives, representing twelve groups and three external suppliers. They met three times weekly to work on the shifter problem, which necessitated revisiting previously "resolved" issues regarding the design of the shifter, the console, and the Styling office's artistic objectives. The team met for approximately four and a half months. During this period, I observed twenty-five formal meetings and conducted numerous informal consultations and interviews with various members of the team. In addition to engineers responsible for the manual shifter (Powertrain) and the console (Interior), the cross-functional team involved engineers and designers responsible for a brake sled, on which the shifter assembly was to be mounted (Chassis); the heat duct below the shifter assembly (Interior); the assembly's noise, vibration, and harshness, or NVH, characteristics (Vehicle Development); the move-macro model of the shifter's movement (Chassis); the G5's overall cost, weight, and schedule (Program Management); and individuals outside of engineering part design responsible for the aesthetic appearance of the shifter and console (Styling); product assembly at the plant (Manufacturing); and three different suppliers responsible for outsourced design work.

The first crunch team meetings I attended clearly indicated that the team was struggling. The team's designers had to make sense of the confusing move-macro simulation output, reconcile this with the 3-D design software, and offer viable design solutions. The difficulty the team experienced interpreting and addressing the move macro undermined the prevailing routines that typically guided the design process to a predictable solution. One crunch team engineer referred to the situation as a "huge ball of yarn." A designer characterized the initial crunch team work as "the most confused project I've seen in thirty years." Even as the move macro disrupted a part of the design process, the larger design routine moved forward while the crunch team struggled to get this part of the design work back on track.

For the first four weeks of my observations, the crunch team was characterized by an absence of mutual intelligibility, and the feasibility and enlistment that might be expected to follow, and instead was characterized by misunderstandings, confusion, discord, and intense frustration. Many of the team's struggles were the result of what I refer to as "slippage," or failures in communication or coordination of action.[7] While mutual intelligibility relies on a minimally viable exchange of meaning, slippage involves the marked absence of such shared exchanges. Where mutual intelligibility involves attention to the audience and its received message, slippage involves the relative absence of such attention. To understand the entrepreneurial action that later emerged out of the crunch team, we must first examine the initial difficulties the team confronted and the slippage that ensued.

During the first phase in which slippage predominated, I observed twenty-one breakdowns in communication and coordination in twelve meetings over a four-week period.[8] These breakdowns in mutual intelligibility included exchanges where the issues raised by one individual were entirely overlooked by a second person in direct response, and often dropped; errors in referencing the details regarding basic design changes voiced by others minutes earlier; related failures to correct obvious mistakes in referencing existing design parameters by other team members; and errors in describing basic features of the move macro.[9] Related slippage issues included disagreements over whether proposals were sent, delayed distribution of critical updates to the move-macro program, and out-of-date information being sent to a supplier. One slippage example was described by an engineer who regularly attended the crunch team meetings:

> For a week and a half [Interior] has been asking for [Powertrain's] changes to the console. . . . The [Interior] guys say, show us a proposal [that summarizes your proposed changes]. Paul [in Powertrain] is saying he gave [Interior] a proposal last week—and [Interior] say[s] no. As of Wednesday, Michael and Eric [in Interior] haven't received a CAD model.

In my crunch team fieldwork, I observed many of these episodes. Such breakdowns were either overlooked or ignored in the team's routines, which repeatedly failed to yield progress toward identifying and solving the manual shifter problem.

At the heart of the crunch team's difficulties stood the move-macro simulation, which had generated the problem finding that necessitated the team's formation. It was that problem that now had to be addressed. The simulation, which could only be operated by a small number of designers, who themselves were unclear how to interpret its findings, served as an enormous source of confusion. Although everyone understood the necessity of addressing the move-macro findings, interviews during and after the crunch team effort revealed extensive disagreement about who in senior management dictated compliance with the move-macro findings, what "compliance" meant, whether the model's assumptions and analysis corresponded with reality, and who could operate or legitimately interpret the move macro. Despite indications from several senior managers that the move macro was a relatively straightforward analytical tool, my observations and interviews with designers and engineers made it clear that the use and interpretation of the move macro posed a considerable challenge.

The failure to appreciate the challenge this technology posed was itself a major stumbling block. A middle manager familiar with the move macro explained that it was more difficult to model the dynamic movements associated with the shifter console than it was for many of the other move-macro applications in the organization:

> We sprung this move macro thing on the organization and never managed the information very well, and we didn't tell them really how to make best practical use of the information. . . . We just gave people way too much data and didn't tell them what to do with it.

In addition, several Powertrain engineers questioned the legitimacy of the move-macro simulation, and whether, for example, the rubber mounts for the engine on which the move macro was based bore a sufficient resemblance to the G5's engine mounts to justify the model's findings. These fundamental questions about the move macro's legitimacy, which introduced a concern around the plausibility of data, helped prevent the crunch team from making progress. This confusion undermined traditional forms of engagement and analysis, and consequently resulted in the absence of actors motivated to lead or reestablish mutual intelligibility within the team.

Phase 2: Contrasting Two Engineering
Efforts to Mobilizing Support for Innovation

Over the course of the manual shifter design effort, two distinct subcases illustrated how combinations of knowledge articulation and brokerage within the NewCar design routine contributed to the success or failure of various innovation efforts. Phase 2 contrasts Eric's failure to gain support for a proposed product innovation with Henry's successful advocacy for an innovative design change to the manual shifter.

Eric's Failed Advocacy for the "Whalebone"

Eric was an engineer from NewCar's Interior division. Two weeks into my observation, he first proposed a daring departure from traditional design that he believed would solve the shifter problem. Using a hand-drawn sketch, Eric proposed replacing the standard hard-molded plastic console with a flexible, soft foam wall supported by a spine of embedded plastic teeth that would flex when hit by any extreme shifter movements. Eric dubbed the idea the "whalebone" due to its skeletal row of teeth (Figure 4.4).

When he introduced the whalebone to the Styling group at a high-visibility meeting in Styling's exotic, glass-walled offices, Eric got a strong indication

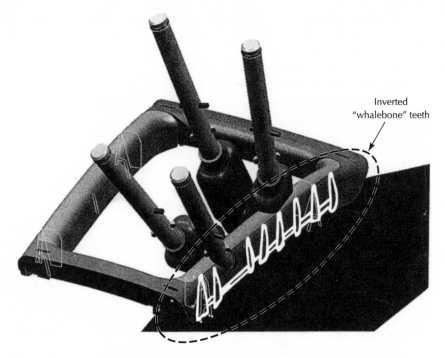

Figure 4.4. Eric's Sketch of the Whalebone
SOURCE: Unofficial figure from defunct automotive firm, 1988.

of support from Rick, the influential lead stylist. But Eric was unable to convert this initial backing into broader support. Over the next several meetings and related communications, he failed to articulate the whalebone approach to make it useful and relevant to the influential Powertrain middle managers. His challenge was exacerbated by the prevailing slippage that characterized much of the crunch team's communications. Eric passively put the whalebone idea in play for other parties to assess, rather than actively pursuing mutual intelligibility or attempting to knit together a coalition of support. Eric's failure to develop the whalebone proposal in ways that addressed various stakeholder concerns undermined his efforts, and meant that he never came close to establishing a base of feasibility.

Eric's failure to generate mutual intelligibility around the whalebone was exacerbated by Alpha, an external supplier assigned with providing a digital rendering of the whalebone idea, who proved unable to capture the essential whalebone concept. By substituting hard plastic for foam, Alpha's design failed to create the crucial flexibility that Eric intended for the console walls, seriously compromising the idea's viability. The whalebone idea was finally ruled out by a senior Powertrain manager, Jim, while he was preparing for a follow-up EDRM presentation:

CRUNCH TEAM MEMBER: What about the whalebone proposal?
ERIC: It fell off the table somewhere. . . .
JIM: I didn't sign off on it. It's not acceptable.
ERIC: I didn't know that. That's the first I heard of it.

Eric's comments, "It fell off the table somewhere" and "I didn't know that; that's the first I heard of it," reflect his passive stance toward both his proposal and its perception by key stakeholders. Eric's unsuccessful pursuit of the whalebone idea derived from his failure to articulate the idea sufficiently to different stakeholders. He never established mutual intelligibility as to the minimum design requirements for the whalebone to be executed. As a result, Eric's solution was never seen as feasible or to represent an attractive opportunity, and the whalebone concept was never seriously pursued (persuasion). As a broker, he failed to reach out to Powertrain or build a cluster of support among various stakeholders in Interior (his own group), Styling, or Program Management. His failure to advocate left the idea socially inert and outside of the collaborative design process.

Henry's Successful Advocacy for the "Double Boot"

Henry's Ongoing NVH Promotion

In marked contrast to Eric's failure to gain support for his proposed product innovation, a second engineer, Henry, successfully advocated for an innovative design change to the manual shifter. Henry was an amiable engineer with a mandate to reduce noise, vibration, and harshness (NVH) for various AllCar vehicles. Henry, along with NVH, "belonged" to Vehicle Development, one of the engineering departments which helped coordinate the integration of different parts. He had a deep base of engineering and organizational knowledge drawn from thirty-four years of experience in the automotive industry, twenty-five of which had been devoted to NVH-related issues. His boss described Henry's job as a "mad scientist role" and one that gave Henry "full license" to pursue NVH-related issues. Yet that license to advocate did not come with the authority to demand compliance; rather, Henry was authorized only to engage in design effort where he felt NVH issues might be at stake. In the G5's manual shifter redesign effort, Henry saw an opportunity to reengineer the subassembly with NVH in mind. His aggressive engagement of this issue stood in stark contrast to Eric's passive advocacy.

Henry's pro-NVH interventions tended to be viewed skeptically by part design engineers, who viewed the whole NVH topic as a distraction at best, because it involved more-abstract objectives, difficult-to-measure outcomes, and labor-intensive, collaborative efforts, as opposed to the more tangible task of designing parts for which an engineer could be responsible and take credit. The challenge of enlisting part engineers in NVH-inspired design decisions was summarized in a sign Henry posted on his cubicle wall:

NVH-Related Decisions Made Easy
Check One: For use by decision-makers that are cost-conscious and/or over-loaded with more important work. Useful for Product Planning, Program Management and assorted design departments.

1. It only hurts a little, so we'll do it.

2. It only helps a little, so we won't do it.

3. It definitely helps, but we don't believe all those things are really necessary. Go back and tell us which ones really work.

The sign's opening remark regarding stakeholders worried about cost or "more important work" suggests that Henry was quite aware of his audience's perspective, a key to adjusting his articulation to that audience. He understood that for engineers NVH often took a back seat to more concrete design issues.[10] The next three points poke fun at the rationalizations engineers employed to dodge requests to make design adjustments that would reduce NVH. As Paul, the Powertrain engineer and crunch team leader, told me privately, "Some say [NVH is] a crock of shit. . . . In the past . . . these guys were ignored. Then noise was generated, and then you'd have to generate a Band-Aid." With this resistance in mind, Henry used the disruption that occasioned the formation of the crunch team as an opportunity to advocate for a change in the rubber boot that surrounded the shifter and sealed off road noise from the interior cabin.

Whether the G5 vehicle had an automatic or manual shifter, a hole in the automobile's floor was necessary to connect the transmission underneath the car to the shifter inside the car. The current G5 design, which called for a single rubber seal around the manual shifter, was Henry's key concern. The manual transmission required a much larger hole in the floor of the automobile than the automatic transmission did. Henry believed a newly engineered part with a double rubber seal, which he referred to as a "double boot" (Figure 4.5), would provide an important additional barrier to external road noise, and therefore greatly improve the G5 manual transmission driving experience. This double

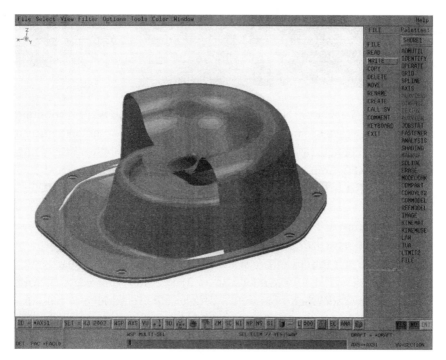

Figure 4.5.   CAD Illustration of the Double Boot
SOURCE: Unofficial figure from defunct automotive firm, 1988.

seal design had never been attempted at AllCar and therefore represented a substantial innovation. While Henry's innovation did not address the core interference problem, he used the disruption in the manual shifter design process as an opportunity to advocate for the double boot. Henry ultimately convinced the crunch team to adopt a manual transmission double boot through skillful knowledge articulation and brokerage.

Henry's Knowledge Articulation: Communicative Dimensions in Action

Henry had a keen sense of the importance of timing, which is evident in the following remark:

> One of the guys who was NVH king around here for a few years kind of drew a curve of the right window for putting NVH in the design. It's about a three-week period, and you've got to identify it, you got to be there. Not only that, but that right three-week period is different for different parts of the car, and if you're too early, then nobody wants to talk to you or nobody has even thought about it; if you're too late, well, it's cast in stone.

This awareness helped Henry understand that he needed to advocate for the double boot at the time that he did. Using the crunch team's formation as his window, he launched a multipronged effort to articulate the importance of NVH and mobilize support of key NVH design changes.

NVH Framing Presentation   One of Henry's main objectives was to reposition NVH from a "crock of shit" to a worthy design objective that should transcend organizational boundaries. To accomplish this, he had to establish a base of mutual intelligibility from which he could begin to establish feasibility and buy-in among key stakeholders. Throughout the crunch team initiative, therefore, Henry tracked down key stakeholders so that he could meet with them and present an informal NVH presentation he had developed over two decades. The presentation, which provided a systematic overview of NVH, was given to individuals and groups of key stakeholders outside of the regular crunch team meetings, typically at their desks. It consisted of forty-one pages of charts and hand-drawn stick figures, as well as decades-old acoustical test results that stressed the three ways acoustic materials could prevent noise: as barrier, absorber, or damper. A series of eighteen stick-figure drawings illustrated essential NVH principles, making the concepts more accessible to the other engineers and designers.

Two of these drawings (Figures 4.6 and 4.7) were used to show good and bad solutions, from an NVH perspective. The first showed a thick barrier preventing waves from moving from one side of a divided tank to the other, with the result that the stick figure in the receiving side of the tank, in water up to his neck, could wear a smile on his face. The stick figure in the other drawing, this one with a hole in the barrier, had a worried look on his face because the waves threatened to swamp him. This drawing was titled "Barrier with Hole": a bad situation from an NVH perspective.

Over time, Henry's articulation of these often overlooked concepts of sound transmission, which all of NewCar's engineers had learned in college, legitimized his NVH-enhancing objectives and designs, thus overcoming an anti-NVH bias that many engineers harbored. Henry's low-key, respectful presentations, which I observed on three occasions, created mutual intelligibility through a familiarizing move and alleviated the engineers' anxiety about understanding the physics associated with NVH. During one of Henry's informal presentations, Paul, the crunch team leader, admitted that he had "taken this stuff in school but hated it." These presentations also repositioned Henry from being perceived as a marginal NVH advocate with whom the design engineers

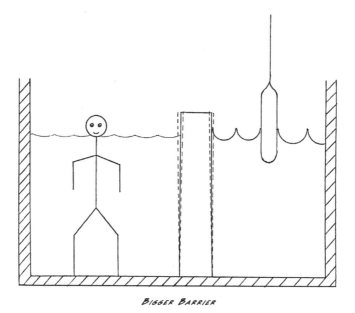

BIGGER BARRIER

Figure 4.6. Henry's NVH Sketch I

SOURCE: Unofficial figure from defunct automotive firm, 1988.

BARRIER WITH HOLE

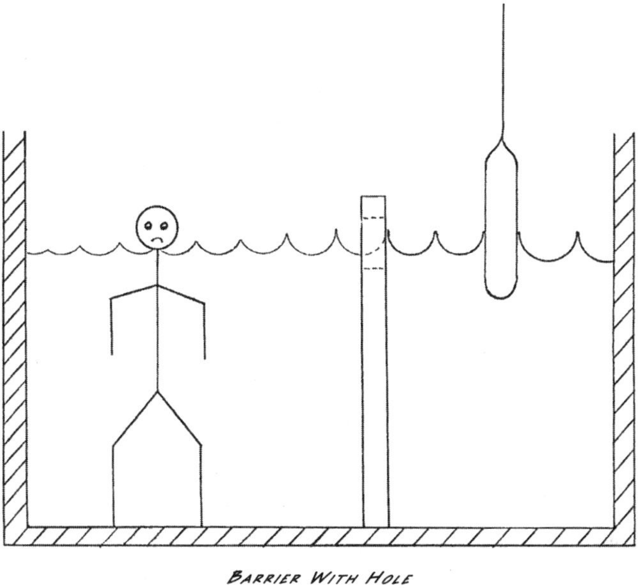

Figure 4.7. Henry's NVH Sketch II

SOURCE: Unofficial figure from defunct automotive firm, 1988.

had little contact, to a familiar, professorial nonpartisan pursuing the highest-quality design.

The presentations elevated NVH by demonstrating its potential to advance the general cause of good engineering, as opposed to the contentious design disputes that otherwise characterized the crunch team. In the process, Henry also created or strengthened ties with various individuals or groups, all of which were crunch team stakeholders. In many cases, side meetings like the ones Henry called to discuss NVH were the only time that task force members from different departments encountered one another, outside of routine crunch team meetings.

Marshaling and Presenting Data   Henry also established mutual intelligibility and feasibility by generating and brokering a second source of NVH-specific knowledge for the crunch team, commissioning a formal set of NVH analyses by which he enacted knowledge transfer and conduit brokerage. He arranged for an outside supplier to perform several lab studies that first compared the sound transmission through a solid steel panel versus an open hole to establish the basic sound transmission problem posed by the manual shifter, and then compared the sound transmission of six different boot designs over the entire sound frequency range. The thirty-seven-page report that resulted provided strong evidence for the merits of a double boot. Henry also prepared a series of precise engineering cross-sections of different boot designs.[11] Even though the double-boot design hadn't been attempted at AllCar, these studies established its legitimacy, in part by demonstrating that other automotive manufacturers had used it. "One thing that helps [when advocating for NVH] is finding out if someone else does it that way," Henry told me. "My [saying] is, 'An ounce of everyone else does it is worth a pound of test data.'" Henry's commissioning of these reports involved not only knowledge creation but also knowledge articulation, in the sense of how he translated the key issues to the crunch team. Henry also assumed the stance of a conduit broker by importing evidence of the boot-manufacturing practices of more than a dozen other car manufacturers into the G5 community.

Confronting the Crunch Team via Defamiliarization   Henry's initial marshaling of data helped to familiarize the crunch team with the idea of NVH-enhancing design changes, and as a result established a foothold of feasibility for the double boot. Nevertheless, the double boot received only minimal support in crunch team meetings. After several meetings, therefore, Henry shifted to a more contentious stance, forcefully asserting the need for a double boot:

HENRY: I have been accused of coming too late [in the design process to recommend design changes]. We've been here every time. I'm concerned about the boot. That design will preclude a new design that we've seen at Toyota. So we need to get the basic design figured out.

BURT (Powertrain middle manager): We can't do that until we get the basic design handled.

CRUNCH TEAM MEMBER: [The boot is] purely cosmetic.

HENRY (*in rebuttal*): The upper boot has a big NVH effect.

BURT: The feasibility of that design has never been developed.

HENRY: That's right, and now I'm asking that it be developed. It needs to be considered as important and not an afterthought.

PAUL: NVH is in the proposal and it is an important consideration.

HENRY: We're on a crash course with what we don't want to happen.

Henry's communication had a defamiliarizing effect, disrupting the team's design routine, its shared understandings of NVH, and its impressions of Henry as a lower-status NVH engineer who could be humored and ultimately put off.

His abrupt, defamiliarizing move constituted a "breach"—that is, an active, intentional move that disrupted routine activities and taken-for-granted meanings (Garfinkel 1963, 1967; Heritage 1984; McFarland 2004). While Henry never directly confronted the inconsistency between Burt's desire to postpone NVH ("until we get the basic design handled") and his criticism that "the feasibility of that design has never been developed," his breach nevertheless put the various part engineers on notice. Henry's more diplomatically introduced yet unsolicited analyses, pro-NVH presentations, and inquiries also periodically interrupted the team's work, loosening the team's taken-for-granted grasp of its routines yet ultimately in a more familiarizing mode. Going forward, Henry combined defamiliarizing moves, which called certain analyses into question, with more diplomatic, familiarizing statements that articulated knowledge more resonant with his various target audiences' interests, and provided the basis for collaboration within and across groups. These defamiliarizing moves also appeared to have the effect of loosening stakeholders' conceptual grasp in a way that made subsequent buy-in more possible.

Translating While Taking the Attitude of the Other[12]   As evidenced by his cubicle sign, Henry's communication often displayed a sophisticated understanding of his target audience. At one meeting, for example, Henry asked a question about the potential for water leakage in the automobile's console, something well outside his typical area of focus. Henry knew that, unlike him, the other engineers weren't typically worried about air leakage; by invoking the threat of *water* leakage, he translated the issues into more relevant terms. When I asked about this exchange after the meeting, Henry explained:

> You must be referring to when I asked about how the lower boot was sealed to the sled. I posed the problem in terms of water leak. If water comes through, other people will solve the problem. If no water gets through, then I'm on my own. I'm worried that the boot is only sealing at the front and the rear. . . . I'm trying to get them good and nervous from a water-leak standpoint so they address the issue.

Henry translated his primary concern into water-leakage terms that he hoped would generate more immediate concern. His inquiry translated his private, complex grasp of the design problem in terms that were more public, simple, and relevant.

Henry wrapped up one of the off-line meetings he had arranged with Powertrain stakeholders in an interesting way, by indicating that he himself once had worked in Powertrain as a part engineer and at that time had argued *against* another major NVH change for which he now was advocating:

> I don't know if Paul told you this, but I was in Powertrain . . . several years ago. On a Hurst hot rod shifter. . . . The [*name of vehicle*] had a terrible motion. I was instrumental in getting the rubber isolator[13] [taken off]. In my "69" [*name of another vehicle*], I took it off and had the lab give me a custom-built part.

By pointing out that he himself was once a Powertrain engineer who had designed parts and also had resisted NVH, Henry established a commonality and trust with key stakeholders, which cleared a path for enlistment. His empathetic observation contrasted with the frank observation he made in a private memo to his boss five days earlier: "I didn't like the look of the proposed rectangular manual shift isolator when I saw it; it looks a bit like that horrible isolator we used on Hurst shifters in the late 1960s."

Overall, Henry displayed a broad base of knowledge articulation skill. Throughout his G5 engagement, he built mutual understanding of NVH, persuaded G5 engineers that a double boot was (at first) not ridiculous and (over time) desirable, and ultimately enlisted key stakeholders. His success was in part the result of selectively moving knowledge from back stage to front stage in meetings and custom NVH presentations; rendering complex NVH ideas more simple and familiar with his stick-figure presentations; applying past stories and knowledge to the current moment in pursuit of desired future states; and demonstrating an ability to both defamiliarize and translate, to establish mutual intelligibility and alignment.

Henry's Brokerage Activity

I've argued in previous chapters that knowledge articulation should be viewed alongside brokerage process as a source of social skill. Henry's articulation moves should be viewed alongside the relational ties he created, cultivated, and brokered through a mix of conduit and *tertius iungens* approaches.

Henry had a more diverse network than other engineers did, which exposed him to a much broader range of data and technological developments. This was a function of his unique cross-functional, coordinative role, as well as years of industry experience.[14] As noted in Chapter 1, even if an open network affords the broker a broader view to engage as a conduit, the broker must still act to identify the idea or opportunity and lift and move it to a second location. Henry's introduction of the double-boot idea into the NewCar context, for example, entailed conduit brokerage, as it actively imported new ideas like the double boot and other technologies in use among the G5's competitors.

Henry's brokerage activity facilitated new coordination among both unconnected and previously connected people within the crunch team. This coordination occurred first through the establishment of dyadic ties and then through the organization of NVH presentations, in which members from a given department (e.g., Interior or Powertrain) assembled to hear his NVH presentation. New coordination also occurred through emerging cross-departmental support for NVH and the double boot that Henry facilitated in crunch team meetings. Henry's compelling pro-NVH stance, along with his carefully observed, nonpartisan, but engaged positioning on other issues, enabled him to gradually enlist the crunch team members' collective support for the double boot. In addition, Henry mobilized key senior managers in his own department to attend meetings where important crunch team decisions would be made. This mobilization activity between individuals who were previously disconnected or uncoordinated constitutes *tertius iungens* brokerage.

By the end of the project, Henry had successfully mobilized broad support for the double boot, persuading the crunch team to use the double-boot design and make another key NVH-enhancing design change. In so doing, he took advantage of the disrupted routine and social order to advocate for NVH. In an interview several months after the conclusion of the manual shifter design effort, I asked the senior Powertrain manager whose group had initially opposed the double boot on the basis of cost, and who had personally killed Eric's whalebone idea, what innovations his group had introduced to the G5. The manager stepped away from his desk and then returned with a G5 double boot, which he proudly placed on the desk in front of me. The double boot had officially graduated from an illegitimate proposal by an outsider to a point of pride for the Powertrain manager.

Phase 3: Contrasting Efforts to Mobilize Support for a Design Solution

The third example and phase involves two designers, Alex and Joe, who ultimately helped resolve the crunch team's confusion and reach a design solution, in part by establishing the double boot's feasibility and attractiveness as a design opportunity (persuasion). As with our previous two examples, this one contrasts the successful knowledge articulation and brokerage moves of Alex and Joe with the efforts of an unsuccessful designer (Sam) who preceded them.

In a typical cross-functional NewCar meeting, engineers responsible for specific parts would sit around a conference table. Each cross-functional team was led by a design engineer, who was expected to lead team discussions and take responsibility for that team's progress. In those meetings, a designer sat at the back of the room at a terminal operating 3-D software that projected the part designs under consideration. Part design engineers relied on designers, the equivalent of digital draftsmen who worked under their supervision, to create and manipulate 3-D, geometrically accurate computer images.[15] According to one engineer's memory, this advanced software replaced what involved positioning identically scaled drawings on transparent velum on top of one another in order to determine fit; and more recently, two-dimensional computer images. The new Virtual Car 3-D technology, actively under development on the G5, put AllCar at the forefront of world automotive design technology.

Sam's Failed Design Advocacy

During the first five weeks of my observations, Sam, an older Powertrain designer, had the main responsibility for the manual shifter design. Sam assumed a traditional designer's role, taking direction from team leader Paul. The crunch team's charge to make sense of the move-macro findings initially fell to Sam. This expanded role involved painstaking identification of the problem moves in the move-macro simulation, determining their design implications in the 3-D design software, and providing guidance on adjustments to the console design that would eliminate the problematic interferences. The difficulty of this assignment, with its associated technical, social, and political demands, was widely under-estimated, and Sam was never able to rise to this intersecting set of challenges. Instead, he bitterly criticized the move macro (calling it "full of shit") and faulted the way the crunch team worked. He also periodically questioned Paul's judgment, sometimes in crunch team meetings.

Sam operated the 3-D software at the crunch team meetings, bringing up various images for the teams as requested. Though he made minor points about the displayed images, and the occasional sarcastic remark, he never proposed a line of analysis or problem-solving approach that would proactively address the team's issues. This passive behavior was not inconsistent with the other designers I observed, but it created a large analytical deficit for the crunch team, which needed insight on the move macro, a deficit that for several weeks went unnoticed and unaddressed. Sam, in short, was pulled into the team's confusion, rather than creating a path out of it. He never developed a sense of how to build mutual intelligibility around the current source of the problem or his proposed solution, or feasibility for that solution. He once observed, "I'm a thorn in everyone's side. A pain in the ass. I'm not helping, but I'm not hurting." Sam never offered an approach or suggested a path toward resolution that others could build on.

Alex and Joe's Successful Design Advocacy

In the fifth week of my observation, the crunch team continued to flounder. The team's progress toward a solution, as well as Henry's advocacy for the double boot, changed dramatically with the arrival of two new designers: Alex and his boss, Joe, from the Interior department. Consistent with the crunch team's prevailing confusion, no one initially recognized their arrival, knew who they were, or understood why they were attending team meetings.[16] As it turned out, Alex and Joe were instrumental in leading the crunch team out of its confusion. They did so by engaging in several knowledge articulation and brokerage moves that resulted in design process innovations and ultimately the resolution of the manual shifter issue.

Alex and Joe's Knowledge Articulation: Communicative Dimensions in Action

Initial Breach of a Crunch Team Meeting   In their third crunch team meeting,[17] Alex and Joe broke with the prevailing routines and confusion, asserting themselves in ways unlike any other designer I had observed at NewCar. Alex, without introducing himself or asking permission, and despite his newcomer status, first cross-examined and then criticized a supplier regarding his seemingly slow timetable for providing a design ("The nineteenth? That's five days!").

One source of the crunch team's confusion was the unclear participation and responsibilities of two suppliers who appeared to be competing for the boot contract at an uncharacteristically late stage. In the same meeting, after a confusing exchange took place between the two suppliers, Joe interrupted and asked, "What just happened here?" Alex followed suit, asking, "Are there two different proposals?"[18] In an effort to figure out the situation, Joe then asked several questions of the second supplier. When Paul, the crunch team leader, began to answer on the supplier's behalf, Joe interrupted him and said, "I want them to answer. *You're* not making the part." The designers' behavior stepped well outside the traditional passive designer role I had observed in the crunch team and other meetings, and constituted a wrenching departure from the prevailing state of slippage. It provided a powerful first push toward building mutual intelligibility at a basic level, establishing a fundamental grasp of similarities and differences in knowledge and interests.

Active Alter Articulation as a Prelude to a Brokered Solution[19] In addition to exposing the misunderstanding that regularly undermined the crunch team's progress, the two designers aggressively pursued mutual intelligibility with the various stakeholders. As they stepped just outside the meeting described above, Alex and Joe laughed over the confusion they had just observed, after which Alex commented, "What a clusterfuck." They then waited just outside of the conference room door, hoping to intercept the two competing suppliers and persuade them to meet with the two designers to establish consistent criteria for their competing designs.[20] This was essentially a short-lived *iungens* brokerage strategy, the first of many manipulations by Alex and Joe of the crunch team network, where they took matters into their own hands to reshape the network.

Realizing they had missed one supplier, who apparently left the conference room by an alternate exit, the two designers sat down to talk with the remaining supplier. During that conversation, Alex aggressively cross-examined this supplier about his preliminary specifications for the part. In contrast to the crunch team norm of failing to ask for, record, or remember other team members' design specifications, Alex leaned forward in his chair directly in front of the captive supplier to urgently extract every relevant detail of the supplier's design. After asking a series of pointed design questions (e.g., "What thickness do you need?" "What kind of clearance do you need here?"), Alex commented:

> My job is to try to figure it out. That's why I ask, "What's the ideal?" "What seems to best fit your needs?" And see what changes to make to existing designs. That's why I desperately need to know what makes you guys happy.

I refer to this questioning as "alter articulation," because it focuses on prompting an alter to articulate his or her knowledge and interests as a prelude to brokerage. Over the next two weeks, Alex and Joe met with each crunch team stakeholder to identify their different design specifications and enter all of the corresponding digital information in the 3-D digital CAD system.[21] Starting with the suppliers, the pair also sought out stakeholders no one else had thought to contact, including Sam (the Powertrain designer who had been relieved of his lead designer responsibilities), the Styling department, and the G5 Chassis engineer, Phil, who was appointed by management to interpret the move macro and decide which interferences needed to be addressed with design changes and which could be ignored.[22]

As an extension of their alter articulation strategy, the designers demanded specifications from each stakeholder's part, so that they could integrate those varied interests into a design solution. Their fast-paced design jargon, peppered with rapid references to a series of design measures, "laid down a marker" for the designers' competence, one that effectively countered the engineers' inherent status advantage. The ability of the two designers to work within both the social and design arenas allowed them to validate, assemble, and review the community's various digital designs in light of the move-macro findings. The designers also effectively set themselves up to broker a design solution across multiple constituencies. What slowly emerged out of this work was mutual intelligibility about how different parts intersected, and possible solutions for how the team might respond to the move-macro problems.

Creating a Two-dimensional Master Cross-section as a Design Solution   The network centrality Alex and Joe earned as a consequence of their aggressive stakeholder outreach was only their first step toward creating an agreement around a design solution. Even as they compiled everyone's individual part designs into one 3-D "study," there was still a lack of clarity on how to resolve the move-macro interferences. The exotic 3-D digital part representations central to the NewCar design process that had dominated the early meetings appeared to obscure the design conflicts and other issues raised by the move macro. In a surprising departure, therefore, Alex and Joe simplified the 3-D CAD images into a series of 2-D cross-sections of the 3-D images. The designers in effect translated the 3-D image into a 2-D digital framework that alleviated much of the crunch team's conflict and confusion by specifying exactly where differences and conflicts existed. Each "master" cross-section integrated two-dimensional slices of all the manual shifter parts, establishing the compatibility of these parts at specific vertical planes in the manual shifter design (Figure 4.8).[23]

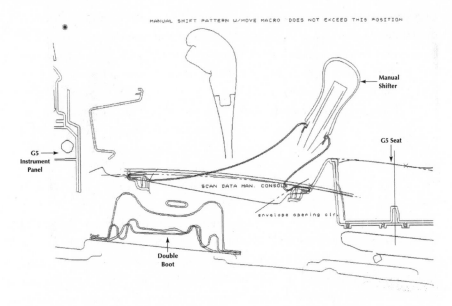

Figure 4.8. One of Alex and Joe's Digital Representations of the Master Cross-section
Source: Unofficial figure from defunct automotive firm, 1988.

The master cross-section illustrates the complex-to-simple dimension of knowledge articulation, by which the creation of an artifact or boundary object (Carlile 2002; Star and Griesemer 1989) allowed Alex and Joe to organize their brokerage strategy. The master cross-section also illustrates knowledge articulation's complex-to-simple dimension, reducing a complex set of 3-D drawings to two dimensions and therefore greater simplicity.[24] The master cross-sections that the two designers constructed allowed them to redirect the disjointed, conflicting engineers toward a resolvable set of decisions that led to a coordinated design solution.

Breaching the Engineering Leadership   Because it departed from the familiar 3-D design routine, Alex and Joe's 2-D, cross-sectional approach required that they assert themselves, compelling the engineers to adopt a markedly different problem-solving routine and power structure. The designers' approach confronted engineers with a temporary loss of control, but it also offered them a way out of their extended confusion. Critical to this approach was the designers' stance as neutral, third-party brokers without any discernible preference for a particular design outcome. Nevertheless, nine days after the two designers' arrival, Joe and Paul had a heated exchange over Joe's attempt to introduce the series of master cross-sections in the crunch team meeting to help reconcile several engineers' designs. In the following exchange, Joe attempts to initiate a dialogue based on the master cross-sections he has projected on the screen:

PAUL (*surprised and slightly confused*): You're designing to our part. . . . No one's talked to me. . . . What's going on here? Talk to me. I'm the engineer on the project.

JOE (*angrily*): I don't care about 3-D [design approaches]. Show me a typical section[25] with how everything fits in.

PAUL: Why are you so upset? You don't even know what's going on here.

JOE: I'm coming up with a typical section. . . . Everything doesn't just burst into 3-D. It starts with a typical section.

We can infer that Paul likely resisted Joe's temporary leadership of the meeting for several overlapping reasons. First, it was unusual for a designer to run any part of a meeting. As an engineer and crunch team leader, this was probably uncomfortable for Paul. Second, as a representative of Powertrain, responsible for the shifter and associated parts that connected the shifter to the transmission below the car, Paul expected the less important Interior department to accommodate the shifter. Paul's comment, "You're designing to our part," certainly reflected his sense of Powertrain's preeminence and the designer's subordinate role. Third, as noted, the 2-D cross-sectional approach constituted a dramatic change from the then-current routine. At the end of this particular meeting, the legitimacy of Alex and Joe's cross-section approach and the potential for the two designers to temporarily lead the team in order to resolve the crunch team's puzzle remained unresolved.

Through their ongoing breaches of crunch team norms, Alex and Joe continued to disrupt the team's action patterns and renegotiate meaning, status, and control. Their knowledge articulation moves indicated an assiduous interest in understanding and articulating different crunch team members' interests as well as a willingness to break from existing practice to confront incongruities and taken-for-granted meanings in order to assert their new design approach.

Alex and Joe's Brokerage Activity

Alex and Joe's advocacy simultaneously relied on a series of brokerage moves that took place alongside their knowledge articulation efforts. The designers first established ties with key players in the design effort and subsequently pressed for collaborative alignment among the various parties around specific design solutions. The master cross-section concept provided a framework for their advocacy, but the success of their efforts hinged on the brokerage moves with which they mobilized support. Overall, the designers' brokerage success was made possible by the intensive, focused attention they paid to stakeholders and the skill with which they constructed mutual intelligibility and a shared understanding of what was feasible and attractive as an opportunity (persuasion). They alternated among different brokerage strategies and stakeholders, selectively relaying critical information (conduit) and keeping people apart (*tertius gaudens*), to achieve their ultimate objective of connecting parties around a design solution made possible by the master cross-section (*tertius iungens*).

Alter Articulation as a Means for Establishing New Stakeholder Ties   As noted, the designers quietly initiated a relationship with Phil, the designated expert for the move macro. Phil was a Chassis engineer whom the crunch team had previously treated as an outsider, even an adversary. Up to that point, the move macro's assumptions had not received much attention. After repeated meetings with Phil, along with extensive experimentation, Alex and Joe began to display a confidence regarding the model's interpretation that I had not previously observed on the crunch team. This enabled them to move more boldly in interpreting the model's findings. In one case, after developing a thorough understanding of the simulation, they simply threw out one problem move that they judged to be impossible in an actual driving situation.

Assembling a Collaborative Network   The designers immediately leveraged the direct ties they created with key stakeholders to coordinate their activity, even though they lacked formal authority to do so. For example, despite ongoing tensions with Powertrain, Alex displayed a sophisticated understanding of Paul's concerns:

> I'm sure Paul doesn't care how we design the cup holders, as long as it doesn't affect [his] shifter. Now he's worried about the height of the cup holders and things in the beginning to make sure we have clearance [for the shifter]. . . . But once we show him . . . where we'll be, then he doesn't concern himself with how we draw the lines as long as we protect what we agreed to.

Being able to show each stakeholder where his part resided within the overall design (whereas in the past there was confusion on this issue) proved a great resource for building trust and dependency. Alex and Joe's efforts extended beyond the different engineering interests, to reflect the politically important aesthetic concerns of the Styling department outside the engineering division. Alex explained:

What happens is each engineer has a concern and he wants his particular product to be the best possible. . . . But at the end of the day, what the stylist describes as the product direction, that theme must be maintained because that's what's been sold to the corporation and that's the direction the corporation wants to take its product. . . . Generally you never [let the lead stylist's] things out of the picture.

The designers' concern with the stylist's position was not trivial. For a design to ultimately succeed, the engineers' designs had to be reconciled not only with each other and with the move macro but also with Styling. Too great a departure from Styling's guidelines could scuttle an extensive amount of engineering work. Recognizing this, the two designers maintained extensive communication with Styling. Rick, the influential stylist highlighted in Chapter 2, corroborated Alex and Joe's coordinative skill, saying about Joe:

He's probably one of the most sensitive engineers[26] when it comes to interior design issues. He's very good at collecting all the respective engineers and getting their minds to work together. . . . And he did that for the G4 program and he's doing that again for us in G5. . . . And a good thing about Joe that's rare—well, not rare, but you're happy when you meet an engineer that really understands what the [Styling] office wants.

In addition to attending to the key Interior and Powertrain stakeholders, the two designers explained the importance of reaching out to Sam, the "overthrown" Powertrain designer, because he understood many of the existing design issues. As Joe commented:

When I went into . . . these meetings, I instantly made an assessment of who were key players and who I should go talk to to get the right information, and who was not full of information. I selected Sam and [his boss]. . . . They are the ones handling the move macro. Those are the people I had to keep in close contact with to get this information. They're the CAD people . . . I need CAD information. I didn't go to the engineer to get his dreams. I went to the CAD people to get the reality of what's coming back as proposals.

Even though Sam proved ill-equipped to solve the crunch team's challenge, the common design language that Alex and Joe shared with the Powertrain designers, most notably Sam, made an informal cross-departmental alliance possible and served as a valuable source of unofficial social information critical to their efforts. By displaying sensitivity to the engineers' issues, the designers cultivated trust across previously contentious boundaries.

Driving Alignment through Selective tertius gaudens Separation and Threats   As described above, Alex and Joe aggressively maintained direct communications with a broad stakeholder base, frequently transferring information among disconnected parties (conduit) but also strategically dropping stakeholders from the dialogue when necessary (*tertius gaudens*) (Davis 2010; Long Lingo and O'Mahony 2010). Alex indicated they had deliberately avoided key stakeholders during a period when they were completing their initial cross-sectional analysis:

> Yeah, we dropped [the Powertrain and Interior engineers] out of the loop, but we got to do some groundwork and build up to a level where they're going to be able to understand where we're coming from with these [master cross-sections]. And now we're at that level. Now we can present this, and at a manager's level they can understand what the obstructions were.

The two designers not only joined people together, but also twisted arms to maintain cooperation. For example, Joe described their response to Powertrain's attempt to change an already resolved design feature, a behavior that had prevented the crunch team from resolving issues in the past:

> I said, ". . . If you guys [in Powertrain] want this, then you and . . . the supplier will have to get together and explain to Styling why now you're destroying everything that was. You're going to have to explain not only to them, but to all the higher echelon [managers] because this thing was high profile.

Here the brokerage involved the threat of acting as the *tertius gaudens* to foment alter-alter conflict. While this brokerage activity didn't actually occur, the threat was a means by which they influenced and enforced agreements with higher-ranking engineers and managers, a twist on the relationship between social network closure and enforcement suggested by Coleman (1988).

Analysis

Alex and Joe ultimately achieved broad enlistment for their master cross-section approach as a frame for organizing the team's problem-solving efforts. At a crunch team meeting nineteen days after Joe's initial arrival, and ten days after his dispute with Paul, Joe initiated a crunch team meeting from the same seat at the computer where, several months earlier, Sam had passively taken directions from engineering. Joe called the meeting to order, selected the CAD images the group viewed, and led the ensuing discussion. By this point, the team no longer questioned Joe's leadership, perhaps the best evidence that he and Alex had successfully articulated their knowledge and made it useful for solving the manual shifter problem, and thereby put an end to the team's confusion.

With their intimate understanding of multiple parties' interests, gained through agential efforts, and their command of the digital design languages that crossed multiple social worlds, the two designers created a novel network and brokerage strategy to drive the crunch team toward a design solution. In contrast to the traditional designer's role characterized by Sam's passive behavior, Alex and Joe, through their skillful articulation of the other stakeholders' design interests, the master cross-section framing, and the mobilization of a temporary network of support, were able to resolve the manual shifter design problem. Alex and Joe, unlike Sam, worked in a tightly linked, cooperative pairing, pooling decades of mechanical and digital drafting experience as well as social knowledge and social support. As with Henry, Alex and Joe's manager gave them more latitude to act entrepreneurially. A subsequent interview with the G5 program manager revealed that he consciously cultivated these designers' entrepreneurial approach when he managed the designers several years earlier, and he made explicit arrangements for his successor to continue supporting that entrepreneurial approach.

Concluding Case Insights

Henry, Alex, and Joe periodically stepped outside, but ultimately worked within NewCar design routines. Their departures from the preestablished design routine were necessary given that their innovation objectives could not be accomplished within the group's established work patterns. Their punctuated departures also involved other subordinate routines such as Henry's stick-figure presentations or Alex and Joe's master cross-section approach. Henry had used the stick-figure presentation before, and I suspect that Alex and Joe had carried out elements of their master cross-section approach in the past, though perhaps not in the full-blown analysis they imposed on the crunch team. The respective breaks from routine introduced by Henry, Alex, and Joe, therefore, most likely involved elements of a preexisting repertoire of action (Tilly 1977) that was adapted and expanded to the situation at hand. In their role as advocates they had to overcome structural and status obstacles to their advocacy. On the other hand, both Henry and the Alex and Joe pairing brought to those situations a distinct agency that involved a broader view of action, a readiness to advocate a change to the existing routine and engage resistance, and a willingness to forge, with the expenditure of considerable energy, a new shared understanding and orientation toward action.

In the engineering community I observed, many had neither the motivation nor the social skill to pursue such intensive coordinative, innovation-generating work. Henry created a shared understanding and grudging respect for NVH out of continuous knowledge articulation along with periodic bursts of brokerage. Similarly, Alex and Joe built a representation of the community, transforming the ability for the group to see their interdependence along with the possibility of solving interferences with concrete steps. As such, both Henry's, and Alex and Joe's work reflected elements of social influence, translation, and synthesis. Both sets of actors, however, never were afforded the stable participation and extended mutual intelligibility (Suchman 1987, 50–51) characteristic of deep knowledge dialogue. Rather, in the absence of such stable participation or sustained intensive mutual dialogue, both Henry and Alex and Joe illustrated the successful pursuit of innovation through provisional knowledge dialogue.

It is interesting to note how rapidly, in the face of certain kinds of disruption, mutual intelligibility and coordination of action can disappear; and of course, how rapidly it might be restored through skillful knowledge articulation and brokerage. Having demonstrated with ethnographic data the dimensions of social skill necessary to get new things done within routines, I now turn to another set of ethnographic cases that illustrate the means by which innovation advocates get new things done in the context of creative projects.

5

Mobilizing to Advance Creative Projects: NewCar's Prototype Parts Purchasing Activity

Having outlined a conceptual framework for how organizations innovate both through creative projects and organizational routines (Chapter 3), and having used ethnographic data to illustrate how brokerage and knowledge articulation account for routine-based innovation (Chapter 4), I now turn to an ethnographic case study in the same automotive setting to illustrate the emergence of creative projects launched in pursuit of innovation. Specifically, this chapter depicts how an automobile manufacturer's prototype parts purchasing (PPP) routine contrasts with two creative projects undertaken to redesign it.

In Chapter 3, based on Strauss's theories of action (Strauss et al. 1985; Strauss 1993), I provided an action framework that distinguished between the trajectories associated with creative projects and organizational routines. What sets creative projects apart from routines is the novel trajectory projection that motivates and guides new, nonroutine action distinct from the repetitive action associated with routines.

I now turn to field observations, again focused on AllCar, NewCar, and the G5, that illustrate this distinction. In particular, I draw on my observations, first of AllCar's prototype parts purchasing (PPP) routine, and then of two initiatives aimed at replacing that routine.[1] I elaborate how *trajectory strategy,* consisting of the trajectory projection and scheme, and *trajectory management,* consisting of knowledge articulation, brokerage activity, and an additional category emerging from my field data, "contingency management," impact the two projects' adoption.

The PPP-related creative projects that I focus on represent two interrelated initiatives. First, a small but diverse group of middle managers from several divisions at AllCar combined efforts to pursue a major change in the corporate-wide PPP routine, which they believed to be costly and inefficient. Second, and concurrently, the G5 program manager, Dan, who was well aware of the AllCar initiative, endeavored to create a new prototype procurement unit to manage the G5 prototype build process. Both PPP-related creative projects were spurred by a vision of how prototype part purchasing at AllCar might be remade, as well as an accompanying scheme regarding how such a redesign might be accomplished.

To lay the groundwork for how the two creative projects emerged, I begin by describing the "cowboy culture" introduced in the Introduction—a culture imprinted within NewCar that gave rise to behaviors through which creative projects were pursued.

"Cowboy Culture" at NewCar

The Mayflower Road Office Building (MROB) that housed NewCar was long known for fostering an entrepreneurial spirit.[2] Originally built in 1927 for the design and manufacture of home appliances, and used during World War II for the manufacture of helicopters, the building over the next seven decades had served as automotive design headquarters for seven different automotive-related companies, most of which were undercapitalized and therefore forced to do more with less. Many of NewCar's employees were quite aware of the building's scrappy, inventive, risk-taking history. In an interview, a former automotive CEO who had started his career at MROB but hadn't set foot in the building for over fifteen years remarked:

> [The company] was not bureaucratic. It had a loosely designed organization chart. It was freer swinging, informal, exciting. It was terrific. . . . We all benefited from the ability to walk into the senior executive's office, sit down, and talk about what you wanted to do, and get it approved. No committees, no position papers, no requests for findings. . . . It was a plutocracy. No one had the money.

That spirit persisted after AllCar's NewCar division occupied the facility. The imperative at NewCar to do more with less led to a string of innovations: the emergence of two entirely new major vehicle categories; the three-dimensional approach to digital design noted in Chapter 2; fundamental breakthroughs in manufacturing and assembly processes; and countless smaller innovations.

NewCar's vice president, Pete, had worked in the building for more than two decades. He had participated in, directed, and cleared the path for numerous innovations, and was proud of the "cowboy spirit" that inspired NewCar employees to take risks. When asked to elaborate on that spirit, he responded:

> [We're] not really overly restrained by the policies and the procedures and the bureaucracy. . . . There's something about the spirit of this facility that kind of exudes confidence and a willingness to get the job done.

When asked about cowboy skills, Pete commented:

> Intuition is a big [part of it]. Using your gut. Using your intuition . . . anticipating. Resourcefulness is a word I like. Resourcefulness . . . finding a way, even including unconventional means, to solve a problem, something that's not already described in your textbook, but finding a way to resolve that. That's a biggie for me. That's what cowboys do, right? I mean, they ride around the range on a horse with a rope and a gun and they built a whole half of our country with their ropes and their guns. I mean, how did they do that? . . . . That's resourcefulness.

Resourcefulness, doing more with less, was a constant challenge for the executives, stylists, and engineers who worked at MROB. As Pete put it:

> We didn't have large resources. We didn't have fancy computers. We didn't have a lot of money. And we had a lot of old-time, hands-on engineers. And resourcefulness was part of that style of engineering. Finding a solution to the problem that didn't involve calling a supplier, writing a purchase order for $300,000, and waiting six months to get something back. We would just run into the shop in the back of the building, grab some iron, cut, weld, you know, mock up whatever we needed to mock up. . . . So we just weren't like the big companies. We had a very small staff of engineers, and the people they hired typically were those who were pretty broad, technically competent engineers who, again, developed a resourcefulness.

"You need zealots," one manager told me. "That's one of our favorite phrases. We need cowboys. We need results. We need spunk, because you're going to run into a lot of people who are naysayers."

There was a shared sense at NewCar that the cowboy culture had been diluted somewhat by the influence of AllCar, the much larger company that acquired NewCar in the previous decade. Nevertheless, in part because of the MROB's location away from corporate headquarters, the cowboy culture still exerted a disproportionate influence within the NewCar division. One reflection of the cowboy culture was the train drawing reproduced in Chapter 2: an image that was proudly displayed in cubicles all over NewCar. It featured a train rounding a bend on tracks that were still being laid as it approached, which symbolized NewCar's culture of figuring things out as they unfolded, instead of in advance. Within the building, there was an informal community of "cowboy" activists who knew each other, had a shared understanding of their past successes, and agreed on the most effective ways to advocate for innovation. At one meeting, the NewCar vice president and several of his managers discussed whether a particular individual was sufficiently "cowboy" to be invited into an aggressive new innovation initiative:

> VICE PRESIDENT: He's a finance guy.
> MANAGER 1: I know him.
> MANAGER 2: Is he wild, wild west?
> MANAGER 1: Not a risk taker, but I sensed that he was comfortable [with what we were proposing].

The tension the cowboys felt between their approach and that of AllCar was clear. In Pete's words:

> The entrepreneurial spirit [involved the approach], "Don't always dot the i's and cross the t's." . . . In a large corporation, procedures are a real common thing. [All-Car] had numerous procedures. And as [NewCar merged with AllCar, the] people didn't like to follow the procedures. Sending a vehicle to the proving grounds, we didn't dot the i's, cross the t's, use the proper procedures; put the proper safety equipment in. It was more [last minute]: "We need a test run tomorrow on this vehicle. Here it is. Run the test." And [the AllCar response would be]: "No, no, no. You have to fill this out. You have to do this. You have to send this along with it. You have to provide this. . . . This is all in the policy." . . . So we really got a reputation of being cowboys and around that [an] unwillingness to just submit ourselves to all the procedures and policies that were in many cases almost militaristic. We thought [Franks, the president] had told us to be [cowboys], by the way.

Another midlevel NewCar manager corroborated the story that Franks, who was destined to become one of the most visible CEOs in the automotive industry, had over a decade earlier explicitly expressed his interest in the cowboys diffusing their willingness to test the bureaucracy and push boundaries:

> [Franks said,] "The only reason we want you guys is because we want you to show everyone at AllCar how to do business more efficiently. If you pick up the procedures manuals and read them and learn all the AllCar procedures, you'll be of no use to me."

While the key features of the cowboy spirit were difficult to pin down, people tended to point to, for example, a desire to pursue innovation that was perceived to be good for the company ("It has to be right to us. . . . What's good for the company is good for me"); acting in the face of opposition by higher-ranking managers ("If my boss won't let me, I'll do it anyway"); doing more with fewer resources than would ordinarily be expected; and a willingness to take risks on behalf of individuals and the company. It is important to keep this cowboy culture in mind as an impetus for the two creative projects explored below. But first I turn to the PPP routine, the major corporate-wide routine whose major problems the two creative projects described further below sought to correct.

Preexisting Prototype Parts Purchasing Activity
Description of the PPP Routine

The PPP routine was a well-defined process for the procurement of prototype parts for assembly into successive generations of prototype vehicles, which led ultimately to a high-volume production vehicle for retail sale. The PPP routine, according to Dan, the NewCar G5 program manager, involved "getting the right parts, to the right [prototype] vehicle, at the right time." Because all design engineers made periodic changes to their parts, the PPP routine could be characterized as a routine for managing modest, relatively predictable part changes.

The core PPP routine involved an engineering request for a part or part change, the solicitation of one or more supplier quotes, the generation of a purchase requisition for prototype parts on a one-time or ongoing basis, approval of the requisition by engineering and finance, the generation of a supplier purchase order, receipt of parts, and supplier payment. The engineering design and testing process yielded a succession of part changes, often anticipated by the purchase requisition. The PPP routine spanned four communities: the part design engineers, the suppliers who manufactured those parts, the purchasing and operations staff, and the prototype build units.

The PPP routine intersected with two other NewCar routines: the engineering design routine and the prototype build routine. The engineering design routine engaged hundreds of engineers whose part designs drove the need to procure parts for a series of prototype builds. In the prototype build routine, prototype vehicles were assembled to provide multiple rounds of feedback on overall performance, as well as on how prototype parts fit together and would be assembled in the final manufacturing stage.

Each engineering division had a dedicated purchasing unit. Design engineers worked directly with part suppliers to generate minor part revisions, but were expected to work through the purchasing units to get approval for purchase orders and major departures from established purchase orders, and to coordinate the delivery of updated prototype parts. Engineers were under pressure to deliver the best possible part in time for a given prototype build, whereas the purchasing unit was charged with supplier costs and adhering to a predictable schedule. The attention that design engineers and their managers paid to costs, timing, and PPP procedures varied widely from division to division.

One unique approach to these arrangements was a PPP unit run by a manager named Craig. His unit focused exclusively on the production of sheet metal and parts related to the automobile's steel outer body. Craig's team maintained relationships with approximately 105 suppliers and actively controlled the engineers' latitude to go directly to those suppliers, as well as suppliers' attempts to work outside of preestablished purchase orders. Craig's skill at coordinating between the engineers and the part suppliers was legendary. With four filing cabinets, five pages of supplier phone numbers taped to his desk, five support staff, and no computer, Craig managed the acquisition of body prototype parts for a half dozen vehicles a year across multiple engineering divisions. Craig was able to pull off this feat in large part thanks to his decades of experience, which enabled him to locate the most appropriate vendors, detect when design engineers or suppliers were making unauthorized changes, and enforce compliance from both.

Trajectory Strategy of the PPP Routine

Trajectory Projection   The trajectory projection, a vision of an expected outcome for the PPP routine, based on past executions of the routine, involved an expectation of a simple progression of handoffs, beginning with an engineer's request for a part or part change and concluding with the receipt of the part and supplier payment. The expected course of interaction was framed around previous repetitions of the established routine. In Dan's words, "Does a [PPP] process exist? Are there clearly defined, clearly institutionalized steps: 1–2–3–4–5–6–7, zigzag left, right, up, down? . . . Yes." Extensive discussions during a three-day retreat convened to chart, diagnose, and consider redesigning the PPP routine (see "The AllCar Creative Project" section below) confirmed the existence of a broadly shared understanding of the routine. This shared understanding was reflected in a high-level process chart developed during the retreat (see Figure 5.1). Retreat discussions surfaced a tacitly held routine that

Purpose: To supply correct new and modified prototype parts on time.

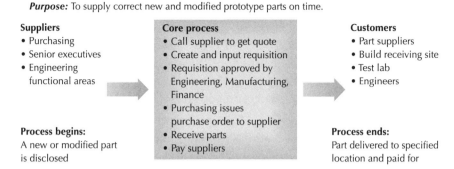

Suppliers	**Core process**	**Customers**
• Purchasing	• Call supplier to get quote	• Part suppliers
• Senior executives	• Create and input requisition	• Build receiving site
• Engineering functional areas	• Requisition approved by Engineering, Manufacturing, Finance	• Test lab
	• Purchasing issues purchase order to supplier	• Engineers
Process begins:	• Receive parts	**Process ends:**
A new or modified part is disclosed	• Pay suppliers	Part delivered to specified location and paid for

Note: This chart is an adaptation of an exhibit from the PPP retreat's final report.

Figure 5.1.   Prototype Parts Procurement
Source: Obstfeld 2012, 1578.

featured a well-defined beginning, middle, and end; included stable participants with clearly identified areas of responsibility; and reflected the sort of repetition and iteration around parts that the prototype build process demanded. In any given iteration of the PPP routine, the end goal, a part arriving on time for installation on a given prototype vehicle, resembled the end goal associated with previous iterations of the routine. This predictable progression and outcome provided the imagined trajectory path that all subsequent executions of the PPP routine referenced.

Trajectory Scheme   A routine's trajectory scheme (or plan) is more tactical than the trajectory projection. In the case of the PPP routine, it involved a relatively predictable progression of tasks for executing prototype part design and redesign. Once the necessary planning was completed, less conscious attention to process was required to achieve the intended outcome; results from the tests of these prototype vehicles would provide the necessary feedback for engineers to revise their part designs as necessary. For example, despite Craig's coordinative talent, he ultimately adhered to a relatively predictable line of action. According to one prototype build manager:

> [Craig] sees things coming, and he's able to predict with a lot of certainty and much confidence what's happening because . . . usually body engineering designs things pretty much the same way [every time]. . . . On the small hinges, tapping plates, reinforcements . . . those are pretty much the same size and pretty much a standard part in the AllCar releasing system.

Craig's attention was focused on the imperative to perfect specific parts and coordinate their design with neighboring parts. As a result, the PPP trajectory

projection and scheme converged and often receded from conscious attention. Phrased differently, the imagined PPP routine trajectory served both as a
projection of an expected outcome and as a representation that guided action
within the routine (Feldman and Pentland 2003).

Trajectory Management of the PPP Routine

Observations of design engineers requesting parts and the PPP group's ordering of parts revealed the relatively straightforward progression implicit in
the widely held imagined view of the PPP routine. That progression, like the
design routine explored in Chapter 4, can be described primarily in terms of
knowledge articulation and brokerage process.

Knowledge Articulation   As noted, work at AllCar involved continual communication, but the discussion that took place within the PPP routine was about
the execution of the routine and dealing with minor contingencies, rather than
in-depth reflection *about* the process or how a more substantial change might
be achieved. Craig described the exchange that might accompany a typical part
change:

> [An engineer will] send me . . . a letter . . . I'll evaluate the cost and . . . if it looks
> reasonable, I'll sign off on it and send it back to him to go ahead and proceed.
> If it's a large amount of money and a time issue, we may have to run it through
> management. Do you want to have this part one week late with this change? Or
> do you want to spend fifty thousand dollars more now, or do you want to wait
> a few weeks and do it later?

Regarding the persistence of engineers requesting part changes, Craig said:
"They'll always come back with one more, [saying], 'Honest. This is the last
change. Honest. You just get this for me.'" Prototype build manager Tom provided a similar description of the relatively minimal level of deliberation involved with supplier quotes:

> [The supplier] . . . sends Craig an official quotation saying . . . $65,000 for the die
> and $110 per part. Now Craig has to take a look at it and say, . . . "Boy, that's a lot
> of money. . . . I don't know if that's going to fly. Resubmit a new number." The
> guy resubmits a new number. Craig sends him a [purchase order].

Craig's work was occasionally contentious; he indicated that he had broken
many phones in screaming fits with suppliers. Such displays generally reflected
his enforcement of AllCar's process standards. Overall, the PPP routine involved
only a modest level of knowledge articulation: that which was inherent in the

coordination among engineers with minor part updates whose successive part orders were handled by Craig's unit and other PPP units. This modest level of knowledge articulation was sufficient to coordinate minor changes in cooperation with engineers, purchasing, and suppliers. It required constant discussion focused on getting work on track and keeping it there. It rarely involved reflection on the process itself or how it might be improved.

Brokerage Activity   Although the PPP routine involved extensive coordination, the typical trajectory rarely involved connecting people in new combinations. Craig indicated that the suppliers he used for the different build programs he managed were "almost all the same people." The stable network of supplier relationships with which Craig worked was reflected in the supplier list. As Craig put it, "I've been doing this for . . . seventeen, eighteen years and . . . I've been working with [some suppliers] seventeen or eighteen years, [so they] are used to me." Although Craig referred to "babysitting" the occasional new supplier, for the most part the relationships he maintained conformed to an established pattern. That pattern was captured in Tom's account of a hypothetical interaction between Craig and a supplier:

> Those relationships are built up. Craig can call up . . . XYZ supplier [and say], "Hey Fred, I've got a part coming over. I want you to look at it. It will be [on the computer] at two o'clock tomorrow. Pull it up . . . and give me a call back and tell me what you think."

In this manner, Craig would periodically convene a supplier and engineer, but the novelty associated with such coordinative action was minimal, and the need for such coordination usually only arose in the early stages of the larger product development process. The PPP routine thus included few novel combinations but, rather, multiple examples of repetitive coordination along established lines of action. Although there were occasional new actors, their engagement was tightly controlled and typically confined to well-established interaction patterns.

Contingency Management   Recall that contingency management refers to the responses to exceptions or uncertainty, and management of disturbances, occurring in the unfolding action trajectory. There was substantial evidence of a "range of interfering and upsetting contingencies" (Strauss et al. 1985, 19) in the PPP routine's execution that led to divergence from the routine. The extent of these disturbances (or contingencies in Strauss's terminology) was consistent with the conception of adaptive routines described by Feldman and Pentland

(2003): they generally stemmed from an inherent tension between the engineering group, which constantly strove to improve the design of specific parts as well as the overall vehicle, and the PPP unit's imperative to get or keep work on track in the face of minor exceptions. Dan, NewCar's G5 program manager, described the propensity of engineers to continually perfect their part designs:

> With regard to the prototype procurement process, it was a good process; it was an institutionalized process; people understood the process, and for all intents and purposes people wanted to follow it. The problem is the tension against following it was . . . all too great. . . . It was the only time you were really able to prove out the design and development and part process concept before you went to hard tools. If you talk about the order of magnitude, prototype tools may cost twenty, thirty, fifty million dollars; production tools cost one billion dollars. You had to get this shit right and you had to get it right in a really short period of time with the whole world watching.

To get their parts right, engineers initiated many part changes both within and outside preestablished purchasing arrangements.

Although many engineering-initiated changes were minor, Dan indicated that some were more significant and could be driven by outside forces:

> It was not unusual at all, for, I'd say, easily thirty percent of the parts in the car needing to change significantly over the prototype build phase—not driven by the engineers' need to change shit, but driven by part failure at the proving grounds, driven by . . . problems in the early stages of build process development, driven by changes in the market where a senior manager saw something at the Geneva Auto Show and said, "That shit looks cool! Let's do that on our car!"

As this experienced manager noted, the forces driving the PPP routine's trajectory to deviate from its imagined trajectory were considerable: "The whole fucking world was against that process." For all these reasons, according to Dan, purchasing and operations managers held a negative view of engineers, whom they described as "wayward": "[The engineers'] basic tendency was waywardness, and 'You need to keep them sons of bitches under control.'" He indicated that he had seen this perspective expressed by senior nonengineering executives at AllCar's highest levels.

With respect to body parts for the NewCar division and another division, the responsibility for controlling the PPP routine fell to the PPP manager, Craig. Craig's success hinged on his ability to predict the path of the PPP routine as a result of its repetitiveness. As he noted in reference to an engineering order in front of him:

I . . . go back to my historical data and figure out based on [the type] of part, how many times did it change design during a program, and how much more money do I think I need. So in [this] case . . . since [engineering didn't] put enough money in [their purchase order], I need about 150 percent more than what they think they need to start with, . . . [or] another $216,000.

The PPP routine was subject to frequent minor contingencies created by "wayward" engineers who often presented part changes, sometimes on a last-minute basis, and more seriously by engineers and suppliers who flouted procurement guidelines by making changes without the PPP unit's approval. Such contingencies might involve changes to automotive parts such as hinges, fenders, or hoods, or sudden increases in the preestablished number of prototype vehicles needed for a given prototype build. Craig described design engineers' attempts to order extra parts off schedule:

You're trying to control [engineers] to keep the changes within reason. . . . [An engineer's] job is to make this one bracket. He wants this bracket perfect, so he'll have it, he'll tweak it and tweak it and keep sending the supplier more and more data. . . . "I want this done and this done." It gets the costs out of hand. . . . Everybody does this.

These pressures required the PPP manager to actively manage the number and seriousness of the contingencies in an attempt to achieve an on-time, on-budget prototype build.

The routine's relative predictability provided a frame of reference for Craig's considerable efforts to corral engineers through close controls on both them and their suppliers. Craig imposed supplier requirements, rigorous reporting, and training—with punishments for noncompliance—to increase the chances that people would conform to the established routine. It was not unusual to hear Craig shouting at suppliers or engineers, informing suppliers that they would not get paid because of their failure to follow his procedures, or telling engineers that they needed to document their part requests or risk not getting the parts they needed for an impending prototype build.

Craig described several ways he managed such contingencies to get work back "on track." One way involved "slapping" suppliers who failed to follow AllCar procedures when they responded directly to engineering part change requests without first getting official clearance from the PPP unit:

The suppliers are told by us, . . . "You do not do changes until you are authorized." [An engineer] will run in with "I got to have this." [The suppliers] do it and unless it's covered [under an existing purchase order], they may not get paid.

It usually takes some little thing like giving them a token slap and [them not getting] paid for something to make them realize we're trying to control things. [One engineer] didn't come to me. He actually went to the supplier. So . . . the supplier came to me this morning [and said,] "[The engineer] needed this [part change]." I said [to the supplier,] "I don't care. I'm not going to pay you for [those parts] now. . . . And [you better not delay the upcoming] part delivery."

Another approach, in response to excessive engineering changes, involved contacting the engineers' managers to ask them to rein in their engineering staff:

I just tell [the engineers], "No. We're not going to do it. No." . . . You meet with . . . the senior supervisors to go through [the part changes] their kids wanted to do. . . . What we can and can't do. . . . [A request] may come through [to] throw all the tools away and start over again. I'm not going to do that unless we really have a real problem.

Negotiations with senior engineering management signaled a more salient departure from the PPP routine, a less frequent major contingency that needed to be dealt with more proactively.

Although the PPP routine typically followed a relatively predictable trajectory consistent with the widely held imagined view of the process, its ongoing execution still generated numerous occasions for contingency management. This comprised Craig's efforts to anticipate and suppress those contingencies, and his efforts to control contingencies once they arose. The PPP routine's parameters represented a truce (Nelson and Winter 1982) between the engineering and purchasing communities. The terms of the truce dictated a variety of ad hoc arrangements and an understanding whereby major disruptions were elevated to higher levels of management, and involved more extensive negotiation between the two communities to determine whether such disruptions would be allowed.

Overall, the PPP routine exhibited more knowledge articulation than is suggested by traditional views of routines that emphasize their tacitness (Birnholtz, Cohen, and Hoch 2007). Minor contingencies, in particular, presented occasions for more knowledge articulation and brokerage activity in order to move work forward.

The AllCar Creative Project

The AllCar creative project group's efforts progressed in three stages, beginning with the mobilization of a core group of colleagues who wanted to redesign the routine, the orchestration of a three-day retreat to generate credibility for the group's collective goal, and a subsequent presentation to Ted and other corporate sponsors. Despite his guardedly positive response, Ted left AllCar two weeks later for a major competitor, after which the AllCar creative project ground to a halt.

Trajectory Strategy

Trajectory Projection   Brian, a frontline operations manager from AllCar's GreatCar division who had spent considerable time over the years with the NewCar cowboy contingent that occupied the same building, mobilized the initial AllCar creative project group. When I first spoke with Brian, he was ready to orchestrate a cross-division initiative to redesign the company's entire PPP process:

> AllCar does a poor job of running the [prototype] business. [It has a] one billion dollar budget but is run worse than a mom-and-pop grocery store. . . . I feel like a voice in the wilderness. . . . Each division does it its own way. . . . [The] downstream production systems . . . don't talk to each other. Across NewCar and GreatCar [we need to] develop a new system that will make up for shortcomings. . . . Create a system; drag some people along.

In brief, Brian was convinced that the current PPP routine was deeply flawed and that redesigning it could save the organization tens and possibly hundreds of millions of dollars. Brian's initial trajectory projection involved a high-level account of a substantially remade PPP routine that addressed a critical, largely ignored set of problems with the existing process.

Brian's expression of these problems and exploration of the opportunity for fashioning an innovative solution attracted a core group of individuals of diverse ranks, responsibilities, and interests. In addition to Brian, the initial All-Car creative project group consisted of Carl, a frontline manager from NewCar who had successfully introduced a number of major innovations in other areas; Bob, a high-ranking manager in the prototype build area at headquarters; Alan, a middle-level purchasing manager for GreatCar; Jason, a middle-level finance manager for both NewCar and GreatCar; George, a frontline employee who worked for Brian; Betty, a frontline purchasing manager from a third automotive division; and Mel, a program management employee from a fourth AllCar automotive division. The group agreed with Brian that the existing process was slow, involved excessive paperwork and outdated information systems, and lacked the essential capability for managing suppliers and costs. During meetings over the first two months, the group evaluated the PPP routine, its problems, and the potential for remaking the routine, and developed a shared vision of a redesigned process built around cross-functional staff, with cutting-edge technology serving multiple engineering divisions.

Trajectory Scheme   The AllCar creative project group's trajectory scheme envisioned a series of mobilizing actions intended to generate support and ultimately lead to an official three-day, human resources-sponsored cross-divisional retreat to assess the PPP routine. The group believed that the retreat could be parlayed into senior management authorization to redesign the PPP routine. This scheme evolved continually, alternating among ongoing consideration of multiple advocacy paths and assessment of the group's efforts as they unfolded.

This retreat-based scheme was only one of several possible approaches the group considered. The range of alternatives was made clear in a response to my question, put to the group after the retreat, as to whether it would have been possible to forgo the substantial effort involved in gaining support for and then holding the retreat as a means of getting an audience with senior management. Carl responded:

> I don't know [the senior executive] well enough to know if he believes in this [reengineering] stuff or not. . . . I mean, if we would have gone to him with the same piece and say, "We've got a group formed of Brian and myself and Bob, Alan, Jason, and George and we want to sit down and meet with you to talk to you about this thing we've got," probably not. Now if we talked to [Vice President 1] in advance and we talked to [Vice President 2] in advance and I got buy-in from engineering and Jason got buy-in for [Finance] and all that . . . then we'd have a meeting with all three of [the key vice presidents] together to put on a presentation. And if they all bought in, then we'd ask [the senior executive] or ask [Vice President 1] to put us on [the senior executive's] calendar. There's other roads to take, other options. But [a human resources-sponsored retreat] is the easiest because for us to be able to [schedule Vice President 1] or [Vice President 2] assumes that we have the validity to do that.

Carl's comment reflects the way in which the group continually discussed and evaluated alternative schemes, an aspect of the knowledge articulation found in this creative project and described below. Brian repeatedly emphasized, as part of the initial scheme, the opportunity to access sponsors and resources by opportunistically linking the AllCar creative project to a corporate-wide initiative, the AllCar Development Process (ADP), to reform product development. This led to the conscious choice to include the ADP group on memos and to invite the ADP vice president to kick off the three-day retreat.

The trajectory projection and scheme of the AllCar creative project traced an imagined line of action that was distinct from the ongoing conduct of the PPP routine. Brian's initial projection, elaborated on by the creative project core group, envisioned an outcome and a means of pursuing it that necessitated emergent, interdependent action directed toward a specific goal. While the trajectory scheme evolved over time, the overall trajectory was initiated and pursued in direct relation to a projection of a redesigned process and an idea of how the PPP routine could be transformed, coupled with a consideration of the various ways and means of achieving that vision.

Trajectory Management

"Trajectory management" here refers to the actions taken within the creative project initiatives to remake the PPP routine. The trajectory action displayed in both the AllCar and G5 creative projects can be understood in terms of the three components cited above: knowledge articulation, brokerage, and contingency management. Whereas the PPP routine's occasional contingencies precipitated increased knowledge articulation and sporadic brokerage activity, the creative project groups' continual articulation of goals and schemes led to the initiation of episodes of brokerage activity and an associated need for contingency management.

Knowledge Articulation   Knowledge articulation was quite pronounced in the AllCar creative project as a means for propelling action forward. In Chapter 2, I introduced three levels of knowledge articulation—mutual intelligibility, persuasion, and enlistment—which are of particular relevance here. In the field, the quest for mutual intelligibility took the form of what I refer to as "fixing"; persuasion took the form of "pitching"; and enlistment took the form of "scheming." In the PPP routine and creative projects, I observed variants of these knowledge articulation forms.[3]

   AllCar fixing.   Fixing refers to knowledge articulation focused on diagnosis and problem solving both within the project team and later, beyond the core group's boundaries. The creative project core group's initial meetings focused on articulating its tacit knowledge of the PPP routine. These discussions involved exploration around a whiteboard, as the group tried to pin down the elusive details of the routine, its problems, and potential solutions. Group members used these initial discussions to establish mutual intelligibility around the problem(s) they wanted to address and to better understand the routine, unpacking its often automatic or tacit nature.

The team's second meeting, for example, considered the PPP process in some depth; the memo that followed featured a "process map" that included the process name ("prototype parts purchasing"), its "purpose" ("to supply correct prototype parts on time"), and other categories such as "suppliers/customers," "inputs," "core processes," and "outputs." The memo concluded with the group's preliminary indications of the process problems or "performance gaps," which included "budget overruns," "no defined roles/responsibilities," "lack of conformity across platforms," and "excess time spent chasing supplier parts/payments." The summary memo indicated that the next meeting would further "define examples of current system waste" and requested that participants "come prepared to discuss actual illustrations from your specific areas." About that meeting, Brian commented: "With the process map, we know where we want to get to. I have to do it slowly enough and subtly enough. It's like you're pulling a wagon with a weak cord."

AllCar pitching.   Consistent with the knowledge articulation imperative to persuade, the AllCar creative project group endeavored to present knowledge about the PPP routine's flaws and potential solutions to key stakeholders for the purpose of enlisting them. In its early search for solutions, the AllCar creative project group made repeated references to Craig's PPP unit. The group spent substantial time articulating Craig's unique role. That role ultimately emerged as a metaphor for the coordinative capability that a centralized PPP unit could provide, which the group came to see as a key solution to the problems with the existing process. In the second meeting I observed, Brian remarked, "We have a hell of a database with Craig."

"He's doing all the right things," Alan responded, "but not doing them for all divisions."

In a critical meeting with a skeptical high-ranking manager whose support was essential, Brian quickly embellished on his rationale for a redesigned, centralized purchasing unit by adding, "Like Craig does for sheet metal," a comment that appeared to allay the manager's concerns. The group increasingly relied on references to Craig, both to efficiently articulate the type of coordinative work the group wanted to see a centralized unit handle and to persuade key stakeholders as to the legitimacy of their approach.

AllCar scheming.   Consistent with the knowledge articulation enlistment imperative, scheming refers to knowledge articulation focused on developing and adjusting plans to guide unfolding action. I use the term "scheming" to correspond with the "trajectory scheme" found in the initial trajectory strategy, as scheming refers to ongoing episodes of active, discursive updating of the initial trajectory scheme as the creative project trajectory unfolds. After diagnosing the process, the group engaged in an extended discussion of participation, focusing on which executives might sponsor an official sponsored retreat and who else needed to be recruited to build broad support for and during the retreat. The group pooled its social knowledge to determine the right mix of supportive invitees. One such exchange went as follows:

ALAN: We need engineers; we have none yet.

BRIAN: I can get you all kinds of engineers.

GEORGE: Forward-thinking people like Steven Pichler.

BRIAN: Forward-thinking body guys in NewCar or GreatCar.

JASON: We need more engineering types from [corporate headquarters too].

BRIAN: I walked into Bob [and asked whether there were] guys "like us" at corporate, and he just shook his head [no].

Over time, the group compiled a list of approximately twenty-five employees thought to be supportive of the initiative, whose attendance would provide legitimacy for any solution the group identified.

Brokerage Activity   In marked contrast to the PPP routine, the AllCar creative project provides many examples of the mobilization of support through the initiation of new combinations of people. Again, brokerage activity refers to the activity associated with creating new combinations in the form of introducing parties, creating meetings, and inviting collaborators in order to move a project forward. What is characterized below as triadic linking also applies to many larger gatherings orchestrated by the AllCar creative project group (Simmel 1950). Knowledge articulation and brokerage activity often interacted, with scheming and pitching serving as the means by which novel combinations were conceived or facilitated.

Case 1.   As noted above, Brian pulled together a core group of individuals of diverse rank, responsibilities, and interests. His effective mobilization of interest and support around his vision constituted the inception of the trajectory. The creative project core group's collaboration constituted a novel combination, because it was discrete from any familiar, repetitive forms of interdependent action within the organization.

At the same time, the novelty of this effort was moderated by the fact that several of the core group members already knew one another and had collaborated in the past, if only sporadically, in efforts to change some aspect of the company's operating processes. According to Carl, "We've been talking about it to each other in different changing networks for five, six years now."

Case 2.   At an early stage, Brian determined that he lacked critical support from the purchasing department, the nominal "owner" of the PPP routine. A supportive purchasing manager would generate credibility for the fledgling effort and help gain access to higher-ranking purchasing managers whose support would be critical downstream. To address this omission, Brian enlisted Alan, the GreatCar purchasing manager, by informally educating him and building his trust. Brian did this by assigning George, a new employee experienced in the prototyping process, to support Alan on several projects. Of this triadic linking action, Brian later said, "I took George in to meet Alan. [Alan] had been in the job a couple of months at most. What helped a lot is I put George at his disposal. Alan is a one-man band. It gave me credibility." In a series of meetings with Bob and Brian several months later, according to Brian, Alan proposed redesigning the PPP routine:

> BRIAN: During those series of meetings we raised a bunch of issues on how
>   we buy stuff. A little light went off and [Alan] just said, "Well, maybe
>   we need to do a process redesign on this." . . .
> ME: Had you already had the idea at that point?
> BRIAN: Sure. But you've got to have somebody in purchasing think of it.

When I later asked Alan about how he began his involvement in the AllCar creative project, he indicated, "It just sort of evolved from Brian and Jason; kind of Brian, Jason, and myself." After earning Alan's trust, Brian introduced him to the core group. Alan would later play a key role in the PPP retreat.

Contingency Management   Managing contingencies within the AllCar and G5 creative projects involved directing an unfolding action path, mainly by anticipating or responding to emergent means and ends that reflected the uncertainty the trajectory management continually encountered (Strauss 1993). Although the trajectory scheme provides an action plan for achieving the vision suggested by the trajectory projection, a creative project's trajectory is shaped by activities in anticipation of disturbances (*anticipatory contingency management*) and in response to events that could fundamentally reshape both means and ends (*reactive contingency management*). Where anticipatory contingency management involves efforts to forestall disturbances or surprises and direct trajectory action toward favorable circumstances supportive of the trajectory projection, reactive contingency management involves disturbances and surprises that demand adaptive responses from the project group in order to move action forward in the generally intended direction.

AllCar contingency 1. The AllCar group practiced anticipatory contingency management when it recognized the need to create the appearance of a multifunctional consensus emerging from the retreat. As noted above, prior to the retreat the group spent several hours identifying a multifunctional list of invitees whom they judged would be receptive to their initiative. Toward the end of the retreat's second day, Alan (the purchasing manager Brian had recruited) introduced a seemingly impromptu solution: a centralized purchasing unit to coordinate prototype builds, which was essentially the same solution the AllCar creative project core group had already developed. Brian, who was sitting next to me at the time, leaned over and wrote on a pad, "Not bad for someone who could not spell prototype [four months ago]." Influential AllCar creative project core group members immediately assented to the idea, and the seemingly grassroots idea of a centralized purchasing unit gained the larger group's quick acceptance. The last day of the retreat was devoted to fleshing out the details of the new centralized group.

AllCar contingency 2. Roughly a month after the PPP retreat and three months after the group's initial work, the group achieved its goal of presenting a proposal to senior management, an occasion that called for reactive contingency management. If Ted bought into the unsolicited proposal, the group would be authorized to move forward in redesigning the process. At the meeting, however, high-ranking executives raised critical issues that the core group had not anticipated, including a caveat that any potential reengineering proposal should not assume the availability of new staff and would need to redeploy existing staff from other units. Although Ted was engaged, he proposed a follow-up meeting in several weeks, at which the AllCar creative project group would provide more details about its proposal.

In the following days, the group set about adapting its proposed solutions to accommodate the new constraints raised in the meeting, such as the "no new head count" stipulation. As it revised its plans, the group abandoned many of its original assumptions about access to substantial new resources.

Two weeks after the AllCar creative project group's successful executive presentation and one week before the follow-up meeting, word came that Ted had accepted a position with another major automotive manufacturer. The project that had steadily gained momentum over several months was at least temporarily derailed. Despite Brian's initial shock, he indicated that the effort was "full

speed ahead," and cited several alternative schemes for moving forward. Eight
months later, however, Brian brought to my attention a memo from Bob that
invited seventeen people, including three AllCar creative project group members, to a workshop to review "part procurement and logistics." The stalled All-
Car creative project was apparently morphing into a substantially different form.

The G5 Creative Project

By contrast with the AllCar creative project, the G5 program manager's successful creative project involved recruitment and mobilization both within and
outside of the NewCar division. While enlisting senior management support
within NewCar, he also negotiated with units outside of the division for individuals to staff his new unit, in exchange for operational support of other
prototype builds once the G5's development was completed.

Trajectory Strategy

Trajectory Projection   The G5 creative project was born from a similar sense
of urgency felt by Dan, the G5 program manager, with whom Brian was in
constant conversation. Dan's desire to create a new unit to handle the G5 PPP
routine reflected his sense of the same problems identified by the AllCar group:
"A prototype tooling program . . . for the entire car can be thirty million dollars. . . . [Several years ago] an early build prototype development phase got out
of hand. . . . They ended up four hundred million dollars in the hole." The immediate impetus for his efforts involved the resistance he had encountered in a
recent NewCar executive meeting:

> We had a five-minute item on the [executive meeting] agenda . . . forty min-
> utes later, [the prototype build manager] and I left the room with tufts of hair
> out of the side of our head where the execs had kicked us in it. They came out
> loaded for bear, [telling us,] "You son-of-a-bitches keep telling us that we are
> late . . . [with our part releases] and we keep telling you that our part schedules
> are changing . . . so you need to just change your plan and stop coming in every
> week telling us that we are late to your plan." . . . I came out of there and I told
> [my boss] . . . there's no understanding on how we got off track and no com-
> mensurate process to get us back on track. . . . [My boss] gives me the Toyota
> heavyweight manager lecture about [how] I own the car, I own everything, all
> these other people are resources to me . . . I said, "Okay." I'm going to call him
> on that . . . I'm basically putting together a little team of people . . . I'm going
> to take over the whole fucking build plan. . . . I'm going to take it over, with or
> without [vehicle build management's support].

Dan's account indicates the close connection between his projection of a new end state (i.e., a new prototype procurement team) and a provisional plan for achieving it, the trajectory scheme.

Trajectory Scheme   As noted earlier, Dan wanted to pursue the creation of a G5-specific PPP team within a much shorter time frame than that envisioned by the AllCar creative project group. A week after his contentious meeting with NewCar executives and after a meeting he had with Brian and Carl, Dan's trajectory projection had crystallized:

> Carl and Brian met in my office. . . . It was a conversation of moles . . . I want to manage getting the right parts to the right car at the right time. . . . I'm in negotiations right now with [my boss] to pull a [prototype] parts group together. I've named them. I'm not sure if [the current build manager] is going to allow it, but I will have to work that out. . . . I am going to want to pick up the pace with this G5 gang.

Dan's remarks indicate some of the raw parameters of his scheme: his ultimate objective, the need to stay underground, the need to secure his boss's support, a preliminary list of staff he would have to individually recruit, and the need to anticipate the existing build manager's potential opposition to his effort. Along with Dan's earlier remark, this indicates how the trajectory projection, trajectory scheme, and ensuing action were more intertwined, in part due to his shorter time frame.

Although Dan's G5 creative project featured a trajectory projection and trajectory scheme, his sense of urgency meant that aspects of both the trajectory projection and the scheme were frequently expressed in the same statements, and led more immediately to subsequent trajectory action. Whereas the trajectory projection and scheme receded from conscious attention in the PPP routine, the projection and scheme were continually referenced and updated throughout the AllCar and G5 creative projects.

Trajectory Management
Knowledge Articulation   As the program manager for the G5 automobile, a high-visibility design effort rapidly approaching the prototype phase, Dan had limited time to dedicate to the discussions focused on a new G5 PPP team. He therefore spent substantially less time diagnosing the process ("fixing") and far more time scheming (with Brian and select NewCar executives) and pitching. Dan's pitch, as noted earlier, was getting "the right parts in the right cars at the right time." This succinct statement anchored most of his appeals for support at different levels of the organization.

In both creative projects, knowledge articulation was used not only to make tacit knowledge more explicit but also to make that knowledge *persuasive*, a clear distinction with the knowledge articulation in the PPP routine. The All-Car group was continually reflecting on "what was going on" with respect to their ongoing redesign effort, and "what to do next" to move the initiative forward. It was through knowledge articulation that the initial trajectory projection and scheme were first developed, and through continued knowledge-articulation activities that the group adapted its approach in response to its unfolding interactions. As noted, pitching activity in both creative projects converged on a particular tagline that over time built the creative case with various stakeholders: "like Craig," for the AllCar team; "the right parts in the right cars at the right time" for the G5 effort.

Brokerage Activity   As with Brian's effort, Dan's initial recruitment activity was "underground": "There isn't anyone that's been vocally against it because most people don't know that we're putting this sort of thing together." Dan admitted, though, that there were two executive engineers to whom he did divulge his plans:

> The execs have to be enlisted. I've had a couple off-line conversations with my two advocates . . . you know, just . . . sort of sneaking around with them. I can talk to them about these sorts of infantile things in private and then they can give thoughts [that are] not fully hatched, early in their development, embryonic diagramming, you know. I just bounce things around [with] them.

Dan assembled three triads of support: himself and the two senior NewCar executives referenced above; himself, his boss, and Carl from the AllCar creative project group; and himself, Carl, and Brian. These triads served to consolidate Dan's support and to explore how to organize and create the G5-specific PPP team. Dan's final brokerage activity involved his successful assembly of a six-person PPP team along with a team leader.

Certainly, Dan did not pursue all possible combinations, and not all combinations he pursued were realized. For example, Dan deliberately chose to leave one executive engineer uninformed about his efforts; and Dan's efforts to link up with his counterpart within GreatCar failed:

> [We] said, "What we're doing on G5 is taking Carl [to write software], getting a G5 build manager, starting that process, . . . and you [*the GreatCar program manager*] should sort of link up with us; learn from us and then from what our team

does, you leapfrog to the next level." He basically said, "Bullshit, I don't want to be linked to you guys."

The evolving combinatorial efforts described here contrast markedly with the ongoing coordination that characterizes the conduct of the PPP routine. The creative project trajectories can perhaps best be understood as involving the mobilization of support through sequences of novel brokerage activity, with one combinatorial effort often serving as a basis upon which subsequent combinatorial work could build. The AllCar creative project group displayed both informal and more structured brokerage activity in the way the initial core group was mobilized by Brian, the provisional alignments the group orchestrated through formal meetings or memos authored by the group on behalf of different stakeholders, and sometimes simply through ad hoc or "underground" meetings arranged to address stakeholder concerns.

Finally, although the creative project trajectories were ultimately defined by a series of successfully created combinations, both also involved many proposed combinations that were never pursued, as well as successful combinations of people that the group initiated but that did not advance its efforts toward reaching its shared trajectory projection.

By definition, the actions in trajectories are interdependent. The data here identify a kind of activity not found as much in organizational routines, where the combinations of actors are more repetitive. Because the combination of people in these creative projects is more novel, the participants cannot predict all the interdependencies, compatibilities, and outcomes in advance. Means-ends relationships in such novel combinations are more uncertain. This uncertainty, where much of the coordinative function attributed to routines (Stene 1940) is not worked out in advance, implies a more active engagement of contingency management, which I next explore in greater detail.

Contingency Management   In his pursuit of a G5 PPP team, Dan also had to overcome the constraints posed by a recent company-wide hiring freeze. Illustrative of anticipatory contingency management, Dan was in continual negotiation with several managers to "borrow" staff, sometimes in exchange for a promise to provide PPP services in the future. The contingency management surrounding these efforts is reflected in Dan's description of his negotiations with Bob (the high-ranking operations manager from the AllCar creative project) to arrange the transfer of an experienced prototype build manager, Kurt, to lead his new team:

Each position is a glove, and it just sits there and the gloves are authorized by the organizations. . . . The individual person is the hand that slips into the glove . . . I have identified a hand. It's called Kurt, but I don't have [a] glove to slip him in. So I'm running around trying to find a glove. What Bob is willing to give me is [his employee] Kurt, which is the hand. But he cannot give me the glove because he needs the glove. If I find a position, if I'm able to get a glove to stick Kurt in, then what that does is, Kurt slips out of Bob's glove and comes to work for me. Bob now has a glove. He would use that glove to convert Will [*a contract engineer*] . . . into a hand. And then he would go and slip into that glove.

Two and a half weeks later, Dan indicated to Brian that he was abandoning the above approach and was hiring Will to manage his emerging group:

Will is officially going to be my [manager.] I can't wait for plan A anymore. Bob can't give up Kurt, and I don't have a means of helping Bob. . . . So we're going to plan B. And we basically have convinced [Manager 1] and [his executive engineer] to give us Will as long as we can backfill. You know, so I've got to verify that we can break Dave free out of Jansen's shop and have him do Will's job.

As it turned out, Dan ultimately *was* able to hire Kurt as his PPP team manager. Throughout the period that Dan was assembling the G5 team, he continually orchestrated both anticipatory and reactive contingency management.

The contingency management examples above indicate that responses involved efforts to anticipate and avoid potential obstacles, as well as respond to disturbances and surprises that actors had not foreseen. The actors forged trajectory paths through the management of contingencies, the outcome of which determined subsequent trajectory action. This contrasts with the PPP routine, where contingency management—with both more frequent minor issues and less frequent major ones—ultimately served the function of keeping or getting work back on track rather than changing the underlying routine.

In these creative projects, contingency management often involved subterfuge. Both efforts initially pursued brokerage activity "underground," or hidden from senior management. In addition, the AllCar group concealed from the human resources facilitators their true motives for scheduling the PPP retreat. Group members created a list of invitees that they secretly expected would support their redesign objectives and that would also build legitimacy by creating the perception of multifunctional support. They approached the retreat with the redesign solution "up their collective sleeve." In the G5 project, Dan covertly negotiated for staff and for senior executive support.

Concluding Case Insights

The emergent action of the creative project and the stable repetitiveness of the organizational routines coexist in constant tension. Earlier I described a NewCar-spawned, cowboy-inspired picture of a train, smokestack billowing smoke, rounding a bend in front of tracks that were still being laid. The train symbolized, in the words of one frontline engineer, how the G5 was "constantly pushing the envelope."

The longer history of that image, however, revealed contradictions and tensions. As told to me by one cowboy, the first response by the executives to the train picture was positive. Soon after, however, a second interpretation of the train—this time, a negative one—surfaced when a group working to implement a particular process (the "Plan-in-Advance," or PIA process) took exception to the train drawing:

> Somebody told [the senior executive,] . . . I think it was the [PIA] guys, "that [the train was] bullshit." They thought we were badmouthing the [PIA initiative]. . . . They took [the train] as a slam. . . . So then they said, "They don't get the PIA principles. Because that's not what PIA is all about."

According to one account, the senior executive reversed his initially positive impression of the train and now held a negative impression.

In an important presentation to a foreign affiliate, a high-ranking New-Car executive engineer presented the train picture as a proud reflection of NewCar's innovative risk taking: "We don't have extremely long range planning. The train is a good picture if you want to summarize how we operate." After the executive described the new Plan-in-Advance initiative, however, one foreign executive asked, "Are you going to draw a new [train] picture for this new [PIA] process?" "The opposite is to have the train in the station," the NewCar executive responded, "and lay tracks through the mountains, and the train doesn't leave the station until all the track is laid." The foreign executive responded, "Like our process!" Much laughter ensued, but again, there was evidence of an underlying tension. As much as the train picture served as an effective articulation of the cowboy spirit, it also surfaced ambivalence about the challenges associated with risk taking and innovation.

Ironically, advanced design technology and the accelerated single simultaneous release design process, innovations for which the cowboys were directly responsible, ultimately called forth far more standardized and controlled procedures, which the cowboys abhorred.

In a related irony, the design innovations the cowboys had instigated led to the creation of a new Plan-in-Advance (PIA) department, tasked with driving a more coordinated and disciplined corporate-wide design process. The cowboy culture of doing more with less and at the last minute, necessitated by a chronic lack of resources, now clashed with a new well-resourced department that was tasked with articulating and monitoring ever more elaborated process guidelines. In effect, the sum total of several creative projects associated with accelerating automotive design was culminating in an increasingly controlled routine.

The examples in this chapter have focused on getting new things done through creative projects. In Chapter 6, I take a more in-depth look at social skill and some of its properties in action.

6

A Deeper Examination of Social Skill

Thus far in this volume I have covered the basics of the BKAP model—brokerage, knowledge articulation, and projects—followed by field data illustrating how these variables fit together in realizing both routine- and creative project-based innovation. I now explore the nature of social skill in greater depth. First, I briefly consider how social skill comprises brokerage, knowledge articulation, and project-based spheres. I then draw on a symbolic interactionist perspective to explore the mechanics of social skill more deeply. Next, I address two more "advanced" social skills, perspective articulation and ventriloquation, and finally close with a case illustrating the use of social skill to get new things done, as exhibited by Dan, the G5 program manager whom I introduced in earlier chapters.

As we've seen, the BKAP model, with its emphasis on brokerage process, knowledge articulation, and creative projects, suggests three distinct spheres of competency. Brokerage competency, as discussed in Chapter 1, involves the strategic deployment of the three brokerage orientations (conduit, *tertius gaudens*, and *tertius iungens*) over time and, more broadly, an ability to identify, recruit, and orchestrate participation as a project unfolds. A second, knowledge-articulation competency involves the ability to read and tailor one's communications to one's audience and context.

A relational astuteness no doubt underlies both these competencies. In part, this involves important considerations of trust defined as a "willingness of a party to be vulnerable to the actions of another party based on the expectation that the other will perform a particular action important to the trustor, irrespective of the ability to monitor or control that other party" (Mayer, Davis, and Schoorman, 1995, 712). Within the brokerage sphere, to choose to connect people (i.e., a *tertius iungens* strategy) and as a consequence divulge information about the project creates a vulnerability for the broker where the now-introduced alters could individually defect or collude in competition against the broker. An alter's choice to participate in a project involves a corresponding vulnerability given the uncertainty associated with committing to an emergent project along with increased difficulty to monitor participants' motives and actions. The broker's choices to make introductions and the reciprocal decision of alters to participate, therefore, are both based on trust. An actor's trustworthiness is based in his ability, benevolence, and integrity (Mayer et al. 1995), an arena primarily associated with the relational sphere.

A third, projective competency concerns the ability to conceive and adapt viable, promising creative projects.[1] Only the first two of these competencies, brokerage and knowledge articulation, involve the social skill necessary to get new things done. Yet being able to conceive a promising new thing is different than being able to get it done. This third, projective competence might account for the early success of a Steve Jobs or Larry Ellison in the relative absence of social skill: an entrepreneur might secure support and participation—and ultimately success—on the sheer strength of the project's vision. This competence is outside of the scope of what we are examining here. The relational astuteness that underlies brokerage process and knowledge articulation is the major focus of this chapter.

Let's first take a closer look at knowledge articulation. One's ability to encode a communication has to work hand in hand with the ability to read one's audience, in order to shape the knowledge that is to be articulated and manage relationships. This concomitant process might be labeled either "perspective taking" or "empathy," depending on whether one attends to another's cognitive or emotional perspective, respectively. In theory, the interplay between knowledge articulation and perspective taking would follow a predictable sequence: the entrepreneur would first read an interlocutor's perspective and interests and then encode a message tailored to those interests.[2] In practice, perspective taking and knowledge articulation are intertwined with each other. I refer to this iterative activity as *perspective articulation*, to which we will return shortly.

Dyadic Dynamics of Social Skill

To understand the underlying dynamics of the BKAP model, it is important to have a firm grasp of relational astuteness, with its emphasis on the ability to read and articulate knowledge to a given audience. While treatments of the relational typically begin, by definition, with the interchange or exchange between *two* actors (i.e., *dyadic*), its full implications in a networked and project-based world cannot be fully realized without a deeper appreciation of its social mechanics and a conceptual expansion to accommodate *triadic* social dynamics. This is the theme I will turn to: the triadic exchange of meanings, in which the broker constructs meaning both dyadically and triadically in order to bring three or more parties together. First, however, I lay the necessary groundwork by beginning with the dyad.

As noted in the Introduction, Weick (1979) argues that the dyadic interact constitutes the core coordinative unit of action in organizing. Recall that the double interact is a three-step process: one person communicates a message to a second person, the second person responds, followed by the first person adjusting her original message based on that response. Note that the double interact also captures the fundamental social dynamic underpinning the minimum threshold for a social network tie, in which two people are connected. The social process for an informational tie, for example, might involve a double interact, in which Jack provides information to Jill in some form of minimal exchange ("There is water at the top of the hill that we could fetch with a pail"), Jill acknowledges and reacts to the initial communication ("Excellent!"), and Jack clarifies one or more aspects of his initial communication ("Let's go!").[3]

Weick is not alone in emphasizing the dyadic interact's role in underpinning organizing. Talcott Parsons (1951) asserted a similar formulation, called the "double contingency," as the fundamental social unit by which the expectations and actions of one participant become oriented to the expectations and actions of the other (Vanderstraeten 2002). More important for my argument, Weick's view of the double interact as the foundation of organizing is remarkably similar to what Mead, in the early 1900s, described as the fundamental "social act" underlying the social exchange of meaning:[4]

> Much subtlety has been wasted on the problem of the meaning of meaning. It is not necessary, in attempting to solve this problem, to have recourse to psychical states, for the nature of meaning, as we have seen, is found to be implicit in the structure of the social act, implicit in the relations of its three basic individual

components, namely, in the triadic relation of a gesture of one individual, a response to that gesture by a second individual, and completion of the given social act by the gesture of the first individual.[5] (1934, 81)

It is out of this two-person "conversation of gestures" that meaning and intentionality emerge. For Mead, meaning and self emerge out of the social act of communication involving two or more people. Mead's approach provides a conceptual foundation for microsocial (dyadic) dynamics but can be extended to illuminate how social skill emerges in triadic as well as much broader contexts.

Three basic features of Mead's account of dyadic interaction—role taking, self as object, and imaginative rehearsal with behavioral adaptation—are central to a more fine-grained grasp of social skill and its emergence. Role taking, or "taking the role of the other," refers to the capacity of one individual to read another's gestures or social actions and impute the other's perspective and disposition to act (Mead 1934). As Joas points out, "The concept of 'role' . . . designates the 'pattern' of behavioural expectation; 'taking the role of the other' refers to the anticipation of the behaviour of the other, and not taking over his position in an organized situation" (1997, 118). To avoid misleading connotations with the term "role taking," Cook suggests an alternative phrase used by Mead himself: "taking the attitude of the other" (1993, 79). If we assume Mead's dyadic point of departure, each skilled dyadic actor imagines the other's attitude or disposition to act and makes ongoing, mutual adjustments in his or her conduct in anticipation of the other's interests.

Mead indicates that a crucial feature of "taking the attitude of the other" is the strategic actor's capacity to see herself as an object, or third person, from the perspective of the other with whom she is or will be engaged. Mead indicates, "It is by means of reflexiveness . . . which enable[s] the individual to take the attitude of the other toward himself" (1934, 134). According to Joas, "The individual now makes his behaviour an object of his contemplation and evaluation in a similar manner as the behaviour of his partners in interaction; he sees himself from the perspective of the other" (1997, 118). Consistent with Joas, I refer to this important additional aspect of taking the attitude of the other as "self as object."

A third feature concerns the strategic actor's capacity to imaginatively rehearse "alternative lines of conduct, to visualize their consequences, and choose actions that seem most likely to lead to successful responses from the environment" (Dewey 1930, cited in Turner 1982, 216). This quality of "mind" for

Mead is a process of behavioral adaptation (Turner 1982, 214), through which a strategic actor can contemplate and compare different social actions such as eliciting support or collaboration—or skirting a potential conflict. Such imaginative rehearsals also suggest another way to understand Weick's famous sensemaking dictum, "How can I know what I think until I see what I say?" (1979, 5). According to Weick's (1979) sensemaking theory, organization members interpret their behavior and environment through ongoing reflection about past conversations with others. From Mead's perspective, retrospective sensemaking would presumably also include the thought experiments that precede action alluded to above.

For an astute, reflexive social actor, imaginative rehearsal serves as a form of social hypothesis, whereby that actor continually posits and tests different hypotheses in his or her imagination before engaging in action. Such imaginative rehearsals occur continually as an actor considers many alternative approaches in sequence. Moreover, if a skilled or semiskilled actor is conscious of the assessment he or she is making in any of these three areas, and then holds those assessments up against the experience generated in action, an additional capacity is generated: the ability to *learn* and, therefore, increase in social skill. The astute social actor learns from the success or failure of the selected action based on the results it produces *in the world*. In other words, the emergence of social skill involves an actor's capacity to learn through dynamic calibration and recalibration of perceptions of others' attitudes, imaginative rehearsals of actions, and the feedback presented by the world's response to a given set of actions. Social skill emerges from refining all aspects of this perceptual and strategic apparatus.

Up to this point, I have explored these dyadic steps based on Mead's work, from the perspective of one actor engaging the other. It is important to recognize that these relational moves are actively practiced by both parties to the double interact. The failure to fully appreciate the bidirectional nature and indeterminacy of this interplay is a critique of Mead posed by Luhmann (1995). In a quick reading of Mead, we might be tempted to assume that these three features of the dyadic interaction pertain equally to all actors. I argue, however, that there exist more and less relationally skilled actors.

To briefly illustrate these concepts at a dyadic level, consider an entrepreneur interested in obtaining financial support from a prospective investor. The entrepreneur might approach such an investor with some limited knowledge based on prior exchanges, for example, that suggests the investor is interested in verifying strong consumer demand for a product before investing, and, more idiosyncratically, that she is disinclined to invest in companies led by younger

CEOs. Thinking through an upcoming meeting, the entrepreneur imagines the investor's view of the upcoming meeting (i.e., taking the attitude of the other), the entrepreneur's product, and the entrepreneur himself (i.e., self as object). The entrepreneur considers alternative ways he might "pitch" the investor to establish credibility for himself and his product. He rehearses a joke that pokes fun at his own young age to put the investor at ease, but decides, after playing out the telling of the joke to himself, that the joke may come off as flip and heighten rather than alleviate the investor's concerns. The entrepreneur might forgo his usual untucked shirt and light beard, choosing to tuck and shave before the meeting.

The entrepreneur/investor example pertains to the imaginative reflection that might occur when there is time to formally prepare for a known investor, but also pertains to the social processes involved when the entrepreneur unexpectedly bumps into a new prospective investor coming off the golf course. Without the benefit of a prior relationship or background information, the entrepreneur might resort to a more generalized representation of a generic investor's concerns and, out of necessity, work more quickly through a more constrained set of imaginative rehearsals to arrive at the best possible response.

Of course, not all entrepreneurs go through similar anticipatory steps, achieve similar depths of reflection and social insight, or reach similar levels of social performance and learning. There is considerable variance in actors' capacity to read different audiences, to make adjustments in their actions, and most important, to learn how to increase skilled performances in a broader range of contexts, even when networks are largely similar. Similar patterns of anticipatory action and learning apply in the brokerage sphere. Some actors grow in social skill over time—that is, they learn—whereas others do not. Consistent with this view, Srivastava et al. (2017) find that people who are able to match their linguistic style of communication to that of their colleagues in an organization achieve higher levels of attainment than their counterparts who are less linguistically adaptive.

This emphasis on social skill can be understood as one means by which actors construct the accurate cognition of social network structure. Krackhardt (1990) found that those with an accurate cognition of informal networks, and in particular the informal advice network, were rated as more powerful by others in the network. An underlying question regarding the origins of cognitive accuracy focuses on how an accurate view of relational ties (either one's own or the ties among others in one's network) is assembled. A structural answer would suggest, for example, that people with greater network centrality might have more accurate cognition of the network structure (Freeman and Romney 1987). The process answer would emphasize the accumulation of many astute dyadic interacts, along the lines described here, to allow for clearer perception of people's motivations, interests, and ties to others.

An important stream of social network research has established the importance of self-monitoring, in conjunction with social network structure, in determining performance (Mehra, Kilduff, and Brass 2001). It is worth noting how this stream of research relates to the concept of relational competence. Self-monitoring refers to individuals' "active construction of public selves to achieve social ends" (Gangestad and Snyder 2000, 546). Self-monitoring research has contrasted "chameleon-like high self-monitors" with true-to-themselves low self-monitors. Consistent with self-monitoring, social skill involves a sophisticated sense of the other's interests and the ability to adapt and respond to those interests, but it also involves the ability to maintain an often contradictory grasp of an independent, agential objective (i.e., the projection) that the actor wishes to carry out.

Of note here is the attention to others that the high self-monitor displays, almost at the expense of his or her objectives. Both the high self-monitor and the strategic actor I have described skillfully adapt their communication to their respective audiences. The strategic actors I describe here, however, are invested in a trajectory projection that motivates and guides action while employing social skill to move toward that objective. In contrast, the self-monitor who employs social skill to align with others' interests would appear to do so in the relative absence of independent strategic objectives, hence the "chameleon" description often found in the literature on self-monitoring.

Triadic Dynamics of Social Skill

Mead has used the concepts cited above to explain different aspects of social phenomena, ranging from the genesis of human language, thought, and self-consciousness to the socialization and further social development of individuals who have already acquired these basic capacities (Cook 1993). Indeed, the dyadic looms large in symbolic interactionist approaches. Mead suggests that the social actor functions not only by taking the role of discrete others, but also by developing the capacity to see oneself from the perspective of the entire organized context of the joint activity or the "generalized other" (Mead 1934). Mead suggests that through internal interaction with the generalized other, the individual learns how to operate within society.

The triadic is often overlooked in the leap from Mead's depiction of the dyadic social phenomena to his depiction of the way actors engage and imagine the expectations of larger social entities, through the concept of the generalized other. C. Wright Mills suggested that the generalized other has a more bounded referent: "My conception of the generalized other differs from Mead's in one respect crucial to its usage in the sociology of knowledge. I do not believe (as Mead does) that the generalized other incorporates 'the whole society;' but rather that it stands for selected societal segments" (1939, 672).

I propose that such societal segments often take the form of smaller emergent clusters of three or greater, whose creation an actor first contemplates and then goes about inducing cooperation through engagement of the knowledge articulation and brokerage competencies. The abilities to take the attitude of the other, see one's self as an object from the perspective of such a hypothetical audience, and imaginatively rehearse different initiatives and consider how they would be perceived by the intended audience before acting, can be conceptualized as involving the perspective of an emergent audience that may grow as collective action scales.

With the triadic in mind, let us now return to Chapter 1's *tertius iungens* case of Fred, Gloria, and Joann. Recall that Fred first enlists Gloria and then Joann before bringing them together triadically—an extension of Weick's double interact. In this vignette, Fred pitches Gloria with a version of the project, P_G, framed to appeal to her, and recruits Joann with P_J. These two double interacts lay the groundwork for Fred's attempt to unite Gloria and Joann, in a *tertius iungens* act of connecting, by pitching a new project blending elements of P_G and P_J, or P_{GJ}. If Fred had assessed his strategic situation as comprising two independent dyadic interacts, he would never have been able to address Gloria's and Joann's separate interests with a blended solution. I refer to this extension of Weick's double interact to include three actors engaged in a *iungens*-oriented coordination as a "triadic interact." This highlights the necessity of using a triadic interact versus an aggregation of dyadic interacts. Indeed, Weick (1979) briefly mentions the triad as an alternative fundamental building block of organizing. I argue that this triadic interact is not a trivial addition to the organization theory lexicon but rather a fundamental unit of coordination and scaling for the creative projects associated with organizational growth and emergence. The initiation of triadic coordination, of course, can arise from any node in the triad.

In addition, to capture a more complete view of the strategic instinct necessary to manage the triadic interact, we must speak to its microsocial origins.

With Mead's symbolic interactionist perspective in mind, consider a strategic actor engaging these social moves (i.e., role taking, self as object, imaginative rehearsal, and behavioral adaptation) not only with one other person, but with a generalized other comprising the two alters simultaneously or with a three-person team composed of Fred, Gloria, and Joann. The entrepreneur might cycle through a series of alternative moves she might employ for each of the alters independently, while simultaneously seeking to identify the type of joint action and framing that will induce the simultaneous cooperation between the two parties she seeks to create. The strategic actor likely fashions a hypothetical, bounded generalized other for which she not only frames or adapts her trajectory projection but also contemplates issues of timing and who next might be drawn into an expanding collaboration. For that purpose, she might also entertain a second, larger generalized other of eight willing participants, and a third generalized other consisting of a community of two dozen participants, some willing and perhaps some on the periphery in terms of commitment or frequency of engagement, which may constitute the organization four months from now.

As the strategic actor contemplates adapted trajectory projections and relational moves necessary to acquire the additional support she needs, she likely touches on all these hypothetical futures, and adapts goals and pitches accordingly. In all these cases, she is bringing together actors, resources, and ideas triadically. The skills acquired in these areas are precisely what differentiates the socially skilled actor from the socially ham-fisted. While structure often creates the pipes and prisms (Podolny 2001) by which opportunities are revealed, social skill provides a lens through which to view how certain opportunities are seen, overlooked, or created through reciprocity, rapport, and restraint.

Such considerations of inducing cooperation are consistent with symbolic interactionism, with which Mead's work is commonly associated, and the broader pragmatist school of thought referenced in Chapter 3. The mediating way of thinking is found throughout the American pragmatist tradition (Mailloux 2011). James (1975) referred to the pragmatist approach as that of a "mediator and a reconciler" and "a mediating system." In the absence of absolute power or structural positions of great advantage, there is substantial leverage in the perceptual clarity and actions of socially skilled actors. Where the unskilled are oblivious to the concerns of others—and therefore can make only limited appeal to the interests of the other stakeholders they must enlist—socially skilled actors acquire influence precisely because of that skill.

Putting Perspectives in Play:
Perspective Articulation of Self and Other

Now let's return to the key topic of perspective articulation, introduced above.

The idea of knowledge articulation, successfully adapted to one's audience, converges on Mead's taking the attitude of the other through perspective articulation. *Cognitive perspective taking* (Davis 1983, 1994) typically refers to the ability to recognize, understand, and anticipate the thoughts and behaviors of others. The roots of perspective taking can be found in Mead's (1934) "taking the attitude of the other" and the "generalized other," noted earlier. *Empathy*, by comparison, is an "other-focused emotional response that allows one person to affectively connect with another" (Galinsky et al. 2008, 378). While Galinsky et al. (2008) have shown that perspective taking and empathy can have differential effects in negotiation, I use the term "perspective articulation" here to refer to actual talk (and other representational activity) that endeavors to communicate one's own cognitive and emotional perspective or that of others. Where knowledge articulation might involve a knowledge representation or argument—for example, in favor of adopting a new product feature—that is framed to appeal to another department or stakeholder, perspective articulation would include the depiction of an individual stakeholder's or department's perspective.

The capacity to understand and express the perspective of others constituted one of the most effective articulation behaviors I observed in my fieldwork. This perspective articulation involved attempts to represent the thoughts, feelings, or experiences of individuals or other groups, departments, or professional categories. Before digging deeper into perspective articulation at New-Car, it's worth revisiting the technical and social dimensions of knowledge in play in that context. The *technical* dimension concerns the skill, experience, and practice generated from participation in multiple product development efforts. This knowledge includes the internalized day-to-day experience of designing parts and subassemblies alongside other engineers from the same and different functions, the experience of success and failure associated with different product development efforts, and the understanding of how different product technologies and materials are applied, along with an understanding of existing products, product limitations, and the feasibility of possible changes.

The *social* dimension of knowledge includes the same shared, day-to-day experience of designing parts noted above, with an emphasis on the history, evolution, and interests of different functional areas and the nature of the relationships between these different functions, including points of friction and alignment over time. The social dimension might also concern the various interests with stakes in maintaining or changing the existing situation, and the concerns of individuals regarding different product development efforts that might inform the tactics and timing used to orchestrate interests at different levels within the organization.

Episodes of articulation may emphasize the technical (as in the case of the stylist specifying the correct surface for an armrest, in Chapter 2), the social (as was the case when Dan, the G5 program manager, was enlisting support for his creative project to introduce a new G5 PPP unit, in Chapter 5), or may combine the two in practice. In both product and process innovation efforts, actors draw on their technical and social knowledge to develop and sell design alternatives, continually responding to shifts in the organizational environment and individual support.

Perspective articulation inherently involves social knowledge, and is closely associated with the back-stage to front-stage communicative dimension of knowledge articulation described in Chapter 2. I organize perspective articulation into four basic categories, depending on whether the knowledge being articulated reflects one's own or another's perspective and whether the knowledge referenced is held at a personal or collective level. The four cases of articulation, then, are (1) an individual's articulation of his or her own personal knowledge, (2) perspective articulation of knowledge held in one's own collective, (3) perspective articulation of another individual's personal knowledge, and (4) perspective articulation of another group's collective knowledge (Figure 6.1).

The simplest and least "social" case of perspective articulation, the one that most directly corresponds with Polanyi's (1958) explanation of articulation in his treatment of tacit and explicit knowledge, is an individual's articulation of his or her own personal knowledge. At a collective level, an individual may also articulate the knowledge held by the group or organization to which he or she belongs. Here the social framing of a perspective becomes more salient. Even though it begins with the individual's perspective, it also involves an attempt to represent knowledge of the community with which the person speaking often has a complex relationship. In previous chapters, I made reference to a drawing of a train coming around a bend just as tracks were being laid before it: an image that articulated the just-in-time ethos that characterized the NewCar division's culture. The frequent postings of the picture on cubicle walls and references to the train in normal conversation were evidence that the drawing articulated an experience held widely within the division.

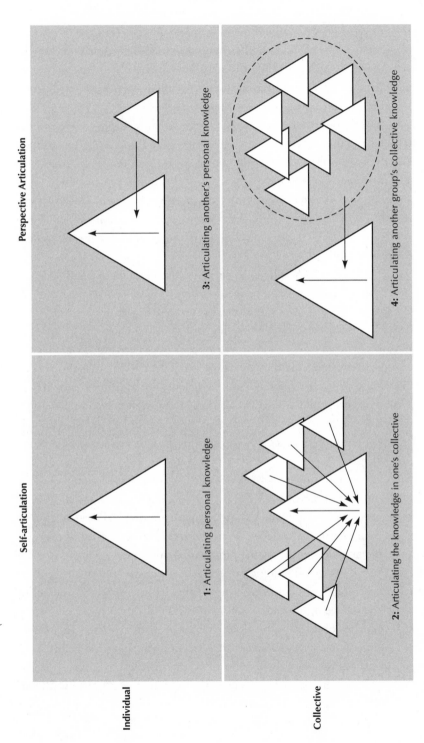

Figure 6.1. A Typology of Articulation

SOURCE: Original figure.

In some cases, an individual's efforts to represent the perspective of his or her community to newcomers or outsiders can reveal new, or forgotten, understanding.[6] As Weick and Roberts write, "When experienced insiders answer the questions of inexperienced newcomers, the insiders themselves are often resocialized. . . . Newcomers are often a pretext for insiders to reconstruct what they knew but forgot. . . . Insiders see what they say to newcomers and discover that they thought more thoughts than they thought they did" (1993, 367).

The third and fourth cases of articulation involve perspective articulation of the "other," in the form of either another individual or a group. After Joe and Alex aggressively collected data about the design interests of the different stakeholders surrounding the manual shifter (as described in Chapter 4), they were in a position to articulate the perspectives of those stakeholders. You may recall Alex's comment:

> I'm sure Paul doesn't care how we design the cup holders, as long as it doesn't affect [his] shifter. Now he's worried about the height of the cup holders and things in the beginning to make sure we have clearance [for the shifter]. . . . But once we show him . . . where we'll be, then he doesn't concern himself with how we draw the lines as long as we protect what we agreed to.

It was that ability to articulate those interests as well as the geometry of the various parts that allowed them to assemble the master cross-section, a two-dimensional digital master cross-section that represented the entire community.

The vignette below illustrates perspective articulation directed at a professional category to which the speaker, Carl, does not belong. Carl, a NewCar employee without a formal engineering education but who has spent nearly two decades working alongside automotive engineers, has constructed a complex, intuitive sense of how engineers work. After a merger with another car manufacturer brings the design philosophies of the two organizations into contact and competition, Carl endeavors to explain how NewCar design engineers work to Aaron, an engineer from the newly merged organization. As he talks, he sketches a sequence of four simple line graphs (Figure 6.2).

CARL: Your work curve—if you're [an engineer who is] supposed to be done at this time, then your work curve should go like this. [*Carl draws a steeply increasing line that flattens out toward the end—Curve 1*] You should get most of your work done and then final up some details. Well, a lot of times engineers will go—like that. [*Carl draws a curve that rises slowly and then increases steeply at the end—Curve 2*] They don't want

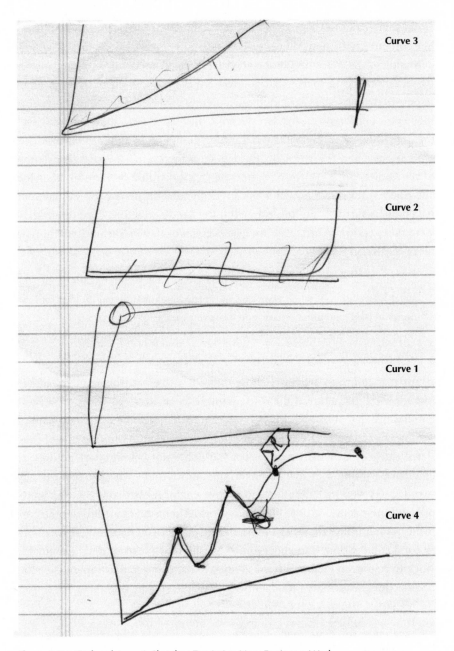

Curve 3

Curve 2

Curve 1

Curve 4

Figure 6.2. Carl and Aaron's Sketches Depicting How Engineers Work

Source: Unofficial figure from defunct automotive firm, 1988.

that visible. As soon as you make it visible to the world that people can see this, now people can see what you're doing. And human beings have a natural thing to—"I don't want to show it to you until it's done." They don't want to be responsible for doing dumb things early. And, it's too bad because half of engineering should be trying new, dumb things. But that's not socially acceptable.

AARON: Additional to your thought, I will make this thought. And that's the problem. Everybody in the process thinks that the engineering process goes [on a straight line] [*Aaron draws curve 3*] . . . [*Carl draws a fourth, bumpy, increasing line—Curve 4—to show how engineering work really progresses. Aaron refers to the bumpy line that Carl has drawn*] . . . [The engineer's customers say,] "Tell me where you are. Tell me you are [here]. [*Aaron points to a spot on the line*] Tell me where you are." . . . The engineer can be here. [*Aaron draws a flag on Curve 4*]

Carl's perspective articulation was so successful that Aaron uses one of Carl's abstract graphs (Figure 6.2, Curve 4) depicting the thought process of New-Car's engineers in order to surface and compare his tacit understanding of the engineers from his own division. Through this process, the voice of the automotive design engineer emerges. Carl first alludes to the performance expectations that certain constituencies place on the engineers. He then uses the hypothetical "voice" of an "engineer" to articulate his empathic understanding of the engineer's tacit experience: "I don't want to show it to you until it's done." Carl's articulation successfully conveys a complex understanding about how engineers work at NewCar, not only with the intent to enlist the other person into his understanding of how engineers work but also to potentially gain his support for a new design initiative. In employing a hypothetical voice of the engineer, Carl's perspective articulation also illustrates a particular variant of perspective articulation called ventriloquation or ventriloquism (Cooren 2010), to which I now turn briefly.

Riffing to Articulate Perspective

Perhaps the most novel form of articulation behavior I uncovered in my study was what I came to refer to as "riffing." Like the more formal linguistic term "ventriloquation," I use the term to refer to oral representations of the voice or lived-in experience of another individual or category of individuals.[7] I call it riffing, a term musicians use to describe instrumental improvisation, because these verbal acts have a spontaneous, performance-like quality that attempts to capture an individual's experience.

I return to Carl, one of the core PPP team members who engaged in some of the most complex and wide-ranging riffs. In the following riff, Carl shifts out of his voice as an operations manager and into the voice of an engineering manager to express a design engineer's preference for design work and resistance to paperwork. In Carl's words (with his riffs in italics):

> That requires a supervisor or manager or an exec to take the nonengineering part of their job seriously. And to be real honest, *You're a manager, you've been doing this stuff for twenty years, and you succeeded by being a good engineer and by doing what your boss wants. Plain and simple. You didn't succeed by always having your bill of materials up to date, by having all your reqs signed, by having all those paperwork things done. And so it's not the highest priority you have. Their job is to solve engineering problems. If you [treat as less important the fact that you] happen to have to sign some reqs on the way to do that, that's [the] mechanics [of] the last six years of engineering schools teaching you're too important to do paperwork....* And that is like that in every engineering organization I've dealt with. [An engineering manager] says, *Hey, what's going on with that door gap flush problem? What's going on with that fender problem? What's going on with that electrical feedback problem?*

This excerpt highlights many characteristics of the riff. First, it is generally embedded in normal conversation. In this case, the riff begins with the words "*You're a manager*" in which Carl, without explanation, presents the perspective of an engineering manager.

As the above quote indicates, riffs can rely on the portrayal of a lived perspective or actually offer dialogue that the subject of the riff might say. Riffing may alternate between these two forms and may also represent more than one individual or category of individual. The portrayal of lived experience may rely on the point of view of the first, second, or third person, or may alternate among them. A riff may also present a lived experience as it currently exists or as it might be in the future; it may also represent a lived experience from the past (i.e., how an engineering manager used to look at the job). In all cases, though, it is the "use of voice" that is important, enabling the speaker to articulate a participant's understanding of the tacit, coordinative work embedded in organizational processes.

The riffing that I characterize as the "use of voice" corresponds loosely with what the Russian linguist Bakhtin referred to as "ventriloquation," or the process where one voice speaks through another voice or voices (Wertsch

1991; Cooren and Sandler 2014). Skillful rendering of a professional jargon can be understood as what Bakhtin referred to as a "social speech type" or "social language." Bakhtin used as illustrations of social languages "social dialects, characteristic group behavior, professional jargons, generic languages, and languages of generations and age groups" (1981, 262). Access to different social languages, and in particular engineering language, was a great potential source of influence in the process innovation discussions I observed. Especially in cross-functional work, individuals who employed large repertoires of social language could construct plausible riffs and compelling worlds that influenced the thinking of multiple audiences.[8]

Riffing varies according to the richness of the qualitative "data" that are featured. I refer to accounts that represent more explicit, available knowledge or only briefly portray the words of a subject as "thin riffs" (after Geertz 1973). Not surprisingly, riffs that effectively convey a deeper or broader social knowledge tend to be more influential. Moreover, those individuals particularly skilled at employing riffs were more influential in determining the shared understanding of the existing process or the merits of potential process changes. In an environment where experience, and in particular design engineering, was essential, skillful riffing created credibility and influence by displaying technical and social knowledge. A skillful riff achieved a certain plausibility or feel of truth and had a tendency to resolve questions and create credibility for the speaker. The persuasiveness of a riff often revolved around an individual's possession of privileged engineering knowledge. This was especially true during the reengineering retreat, where most of the participants did not originate in engineering but had to interact with engineering clients.

Carl, one participant in the PPP core team who had initiated a number of important innovations from a frontline managerial position, was particularly influential at the retreat, in part as a consequence of his engineering experience and broad knowledge, but also in part because of how effectively he used riffing to either advocate a point or frame a conversation. To make the case for why design engineers might favor the coordination that would be provided by the proposed PPP unit, he riffs here on the voice and perspective of a hypothetical NewCar engineer:

> Say I'm a NewCar engineer and I'm building one of these mule cars [*an earlier prototype vehicle*] and somebody is making a custom floor pan . . . so God knows what's going to be on the bottom of the car. But the assumption is that I can route my . . . exhaust pipe around anything. Everybody's [in] Chassis, everybody

does their stuff and then push comes to shove . . . I find out OK, I'm going to
need my exhaust pipe, and I got some information. Does that mean that the
engineer who needs that pipe is going to start calling sources and start checking
out who might be able to bend some pipes for me, or am I going to go to the
central group saying, "Hey, I need a pipe that's bent like this. Here are the X,Y,
Z [*CAD design*] coordinates. Give me a pipe"?

This riff begins with a hypothetical first-person description of an engineering
issue and then contrasts two potential paths the reengineered process might
take. Like the longer riff above, the sophisticated representation of engineering
work (references to "mule cars," "custom floor pan," "X, Y, Z coordinates," and
the engineering design process) articulates a plausible representation of an engineer's lived-in experience, while serving as markers establishing his mastery
of the privileged engineering knowledge.

Carl's remark created a framework for the ensuing conversation, which
took the centralized group as a given. Mel, the other core team member with
extensive engineering knowledge, built on Carl's riff with another riff, displaying similar prowess in articulating the engineering perspective:

And I should have a build plan that says, "I want to build at this rate and this is
my closure freezing point for your profit." What will happen with the exceptions? The exceptions are, I need five extra parts to go off and do testing with. I
need five extra parts to replace parts that failed already. Or I've got some "what
if" thing that I don't even know what the hell to do, but I need help in getting
it done.

In the same two-hour period in which the above riffs were delivered, Carl
offered a surprisingly wide range of other riffs, featuring the voice of a security
guard, if the AllCar parts security system was improved as he proposed ("You
can bring any part you want in here, sir, as long as you pass through my little
part code scanner"); a tough Japanese supplier ("How *dare* you? How *dare* you
delay my build?"); a loading-dock rule enforcer ("If the shipment doesn't meet
our requirements . . . part matches requirement, quantity matches, shipper tags,
[then] the box was not received, [is] put back on the truck and I don't care if
it costs you a million dollars to ship it"); and an engineering operations clerk
with a flawed work process ("If I change this [part] . . . somebody has got to
go through and update ninety-eight individual files. . . . 'Change it on this one.
Change it on this one. Change it on this one'").

The skillful riffs offered by Carl and Mel reflected their broad experience
in multiple professional worlds, through which they acquired a tacit feel for a

variety of different professional environments, perspectives, and languages. The riffs made sense for, taught, and influenced the PPP community.[9] Subsequent interviews revealed that Carl had a background in computers, understood purchasing operations, and despite the lack of an engineering degree had made a point of spending extensive time with engineers over the years. Similarly, Mel had actual experience in design engineering, program management, and operations. He employed a combination of experience and imagination to generate his persuasive renderings of technical knowledge.

Inability to access privileged knowledge or language puts one at a disadvantage, but not one that is insurmountable. Alan, the purchasing person recruited by Brian, lacked the automotive knowledge so valued by the PPP core team and the company at large. Nevertheless, he asserted himself by relying on his pronounced articulation skill. In the core team meetings and the reengineering retreat, Alan continually restated, clarified, and organized the points made by those with more engineering experience. He framed the general direction of the group prior to the reengineering retreat, and was the person who ultimately introduced the idea of a centralized purchasing group to the assembled reengineering retreat.

Riffing proved to be a strikingly prominent articulation skill in *process* innovation efforts, where understanding all the roles and perspectives involved in a process was critical. To the extent that it employed privileged engineering talk or displayed diverse process knowledge, riffing was especially influential in these process meetings, both because it served as a marker of expertise and because it made complex, process-related ideas accessible. Riffing was less of a factor in the *product*-related innovation efforts. When riffing did occur in connection with product design, it tended to emerge in informal discussions outside of "official" design meetings.

The Social Skill Virtuosity of the G5 Program Manager

Let's return to Dan, the program manager over the entire G5 platform and my main informant in the ethnographic study from which this book draws.[10] As several ethnographic interludes in previous chapters have illustrated, Dan's innovation advocacy had enormous impact across AllCar and, arguably, the larger automotive industry. Among other things, he successfully advocated for three-dimensional computer-assisted design (CAD) and the corresponding single, simultaneous release of all part designs on a given vehicle in order to establish overall compatibility (Chapter 2), and for a dramatic reorganization of the prototyping process in the context of the G5 PPP creative project (Chapter 5). In short, Dan displayed impressive competence in each of the projective, broker-age, and knowledge articulation spheres, and was widely seen as destined for the highest ranks of the company.

Given the technically infused nature of the NewCar engineering environ-ment, it was imperative that Dan establish and continually warrant a core level of engineering competence through the skillful use of engineering language (Chapter 2). Beyond this, he had an extraordinary ability to read other people and groups, understand their interests, and tailor his message to those different audiences—threads that ran through all of Dan's advocacy. Toward that end, he was extraordinarily alert to the perspectives of the white executives and mid-dle managers with whom he both routinely collaborated and who frequently sought out his advice.

While I've addressed Dan's talent for getting new things done, that talent takes on a new meaning when one takes into account that the vast majority of his work exchanges took place across a racial divide: he was one of AllCar's few black senior executives. Once I walked in on Dan finishing up a conversation with a white finance manager who visited Dan at his office for advice—a daily, sometimes hourly occurrence in Dan's job as program manager over the entire G5 platform. As I entered his office, Dan concluded the conversation with the remark, "I understood that's not what you meant to say." After the manager left, Dan explained that the manager had sought him out for advice and an expla-nation for why, at the NewCar SVP's staff meeting earlier that day, the SVP (Pete) "took his head off" for suggesting the funds for a particular program he was implementing were not yet available. In explaining the meeting he had just concluded, Dan used perspective articulation to provide me with insight into this manager's finance-based perspective:

> What [the finance manager] had meant to say is, "We're not comfortable with the numbers and not ready to get up in front of [senior] management [to sup-port them." . . . [The finance manager] hadn't thought it through. He took a position his boss [a VP in Finance] took.

Expounding further, Dan articulated, in colorful language, the perspective of the SVP who took the finance manager's "head off":

> [The VP] knows [his executives are] beavering away on how to do the right thing. They need encouragement but they're not cheating, lying sons of bitches. These other groups report to functional VPs [outside of his control] and they *are* lying, cheating sons of bitches. He has to pull his gun and let it off a few times to remind the team who's in charge.

Dan's use of phrases such as "beavering away" and "cheating, lying sons of bitches," and his reference to "pulling a gun" to assert authority, were all typical of the kind of humorous language he constantly inserted into his remarks that facilitated understanding among one or more parties. Dan's observations illuminated how the finance manager had made a misstep but also how the SVP, Pete, was using the incident to make a display of his authority and his division's independence. Dan grasped the complexity of the situation and articulated Pete's perspective sympathetically to the finance manager, who bore the brunt of Pete's public rebuke, while offering advice in part by drawing on how to remedy the situation. He then explained the situation to me (an academic) while adding his own stylistic flourish ("lying, cheating sons of bitches"), which both lightened the story and reflected his own cultural origins. That exchange was representative of countless situations in which Dan brokered between multiple higher-ranking executives or between executives and middle- or front-line employees across NewCar—almost all white men and almost all trying to further various initiatives and their careers. As a consequence of the insightful and forthright way he conducted these transactions, he forged a broad base of trust and rapport across multiple levels of the organization.

Though I make no claims on grasping the complexities of Dan's racial identity, I did become quite close to him and saw the mastery with which he articulated a broad range of perspectives for the purposes of facilitating the various high-impact innovations I have documented in previous chapters. In one meeting, he drew liberally on his Memphis roots as he stressed the importance of middle managers hanging out with senior executives to advance their careers whenever the opportunity presented itself: "So if you have a chance to sit with one of these [senior] folks, drink Colt 45 and smoke doobees, [don't pass it up]. Momma only raised one fool and he lives in Memphis." The remark, which elicited immediate laughter, took me (and I suspect everyone else who heard the remark that day) some time to absorb: first to process, acknowledge, and accept (rather than judge) a reference to an ostensibly "low-brow" aspect of black life; and additional time to unpack the meaning of the fool that lived in Memphis (it meant that Dan, because he didn't live in Memphis, was saying he wasn't a fool).

In a similar vein of humor that I believe was meant to entertain and illustrate while desensitizing his audience to racial differences, Dan often made humorous references to "being from the projects." In one meeting in which I wound up sitting behind him, he turned and said, "Come on in, man. I'm from the projects. People sitting behind me make me nervous." In another exchange, Dan said, "Hey, I grew up in the projects. You can't hurt my feelings. Take your best shot," to which his colleague Brad responded, "Are you going to go through that *again*?" Brad, a casual friend and longtime colleague, had obviously heard the projects reference before and felt comfortable enough to tease him about it. My observations suggested such humor was often successful in diffusing racial tension, including his own.

To help advance our exploration of perspective articulation, I now want to revisit the context in which Dan was manifestly successful: leveraging his projective, brokerage, and relational skill as the G5 program manager. Recall that Dan successfully advocated for the G5 PPP creative project, which involved the invention of a new unit to handle the ordering of prototype parts (Chapter 5). Dan's advocacy for this remaking of the PPP constituted one of several instances of his projective skill. His effectiveness within the executive ranks and across the G5 platform involved a great deal of brokerage skill, by means of which he orchestrated various constellations of support until he ultimately secured support for the unit.

In the following case, I background Dan's talent for projecting new outcomes for the reengineered G5 PPP process, to showcase how he combined relational and brokerage skill to move the G5 creative project forward. Some brief background: Tension had arisen between Kurt, the G5 PPP team leader, and Tim, the prototype build manager in Manufacturing who was responsible for building dozens of G5 prototypes. Both men found themselves engaged in an entirely new, time-compressed prototype build process—and came into conflict. As Dan and I discussed the challenge the situation posed to the success of the G5 PPP creative project, he articulated the perspective with which Kurt, a high-school-educated white male, approached the job, and explained why that perspective was bound to lead to clashes with Tim:

> Kurt, having built cars for years back in the back and having managed the whole issue of parts supply and build issue resolution and making hard decisions, he manages a union shop so he's used to kicking people in the butt and chewing their asses and making them go home because they got alcohol on their breath. And making real time decisions about, "Can I put that ashtray on that car and get it out on time tonight or not? Or do I have to call off the ride and drive?" . . . In his world, in his environment, he's used to getting with the right people, making the hard-hitting decisions . . . Kurt's a shop boy. He's a mechanic. He sort of put away his greasy gloves and tools and eventually became the group leader back in the back but he still has a shop mentality. And he doesn't—he's not into the . . . philosophical . . . this sort of let's smoke some doobees, Dan,

and meditate on why and then based on that we can flesh out the hows and the whats. I mean, he just more or less "Let's get to it" and make stuff happen.

Tensions between the new prototype parts unit—created by Dan and led by Kurt—and the manufacturing personnel had mounted steadily, as temperatures in the non-air-conditioned build area hovered near one hundred degrees. If those tensions were not dealt with quickly, they might well derail the project. In order to keep the new PPP team on track, Dan's mission was to meet with Tim. In going down to the build area, Dan was brokering between the G5 engineering community and the manufacturing division, literally stepping onto Manufacturing's turf. Dan would need to establish a middle ground, in part by "reading" Tim and making adjustments in his own speech and bearing, to first establish a connection and then find a way for Tim and Kurt to work together going forward.

At the outset of the meeting, Dan made small talk with Tim, initially revealing none of his concerns about the current situation with Kurt. Eventually, he brought the conversation around to the subject of the tension between Kurt and Tim. Dan first acknowledged the mistakes Kurt had made:

TIM: Kurt's screwed up in several places.

DAN: Well, Kurt has made some mistakes, and he's been kicked in the head some.

TIM: My point is that there is a lot that we can change and we need to learn from.

DAN: Let me remind you of something you already know. Peterson and Sanders [*two of Tim's managers*] will have your ass if you bring some shit to the plant that doesn't work.

In that last remark, Dan walked a fine line between establishing empathy and toughness. My sense at the time was that the remark probably fit the brasher tone of the manufacturing culture from which Tim came.

The conversation then turned to the fact that Manufacturing had not been provided with the proper tools—a major source of concern for Tim. Dan registered and, through skilled perspective articulation, expressed empathy for Tim's plight:

DAN: You ain't got the time. You've got to bring these engines over and put them in the car.

TIM: They said, "That will cost six to seven hundred dollars." I say that's not my problem.

DAN: You're just trying to build cars.

TIM: They had no idea the special tools they needed.

DAN: If they were over at [headquarters], they would have those tools.

TIM: That's key. Over there they have, "Take this head off here, put this head on here." It takes years to accumulate that level of tools. The sad thing about this facility, it's not equipped. . . .

DAN: It's your responsibility to get the shop ready. . . . One thing, no matter how close you get involved with the plant [*"the plant" is the factory where the G5 will eventually be produced*], the bottom line is that we never got close enough to the plant. We should have had you up there. They walked me through Thursday. . . . We should have had your people up there.

Dan attempted to create trust by pulling on two levers of trustworthiness noted earlier: first establishing competency as an experienced, confident automotive executive; and then engaging in empathetic perspective articulation as he indicated his support of Tim. The difficult situation concluded with a round of teasing and chat that would have led an outside observer to assume that Dan and Tim were good friends, a fairly common outcome for Dan. His effort to diffuse the considerable tension in the build area proved successful, allowing the new prototype build process to go forward.

Dan's complex blend of perspective articulation involved both a cognitive and an emotional perspective of his audience, a critical contributor to his effectiveness in the pursuit of innovation. Given the wide, cross-functional scope of the initiatives Dan pursued, his command of multiple perspectives is even more critical to his success. The examples above provide evidence of his mastery of engineering language along with practical insight into finance, prototyping, manufacturing, and senior executive states of mind. His social insight allowed him to capture the details of the social situations around him with more precision, making adjustments as he went, including the translation of issues, the use of different languages associated with different arenas, and the orchestration of networks.

I began this chapter with reference to the intersections of the projective, brokerage, and knowledge articulation spheres of competence. These spheres present themselves in a (theoretically) logical sequence. First, the projective motivates and directs action toward getting new things done. That sense of the projective, in turn, informs the entrepreneur as to the actors, resources, and contexts that are necessary to move toward the projective goal. A brokerage competence dovetails with the projective, allowing the entrepreneur to consider which actors must be assembled, at what time, and in what manner, to advance the project. Each brokerage calculation in turn quickly implicates knowledge articulation as a means of forging rapport and articulating common interests, as a basis for determining which brokerage orientation(s) would work best. In the real world, of course, these three spheres are intertwined with one another. Together, in their richness and complexity, they constitute a broader framework for understanding and advancing innovation.

To illustrate this point more broadly, I now turn to several applications of the BKAP model to illustrate and underscore the utility of this model for getting new things done.

7

Additional Applications and Implications

In this chapter, I briefly apply the BKAP model of action to a number of important theoretical and empirical puzzles that have been confronted by organization theorists in particular and by social scientists more generally with the goal of illustrating the "Swiss Army-knife" applicability of this construct and inciting further research. While even more applications are possible, I will focus specifically on the applicability of the BKAP model to several mobilization-related issues in entrepreneurship, collective action, and inequality.[1] Given the focus in the foregoing chapters on innovation in business, I lean heavily here on instances in which the model enables users to extend beyond this realm. In order to illustrate the model's broadest possible reach, let's begin far afield, with the emergence of an innovative dance company in early twentieth-century France—a distinctive "start-up" environment indeed.

An Artistic Movement: The Ballets Russes

The Ballets Russes, widely viewed as the most influential dance company of the twentieth century, was created by the Russian impresario Sergei Diaghilev. In his role as impresario, Diaghilev forged collaborations with unique combinations of talented choreographers, composers, designers, and donors. The company, which performed in Europe and the Americas between 1909 and 1929, is credited with revolutionizing dance, revitalizing opera and scenography, and influencing neighboring artistic worlds such as fashion, furniture, and textiles.[2] The young artists whose work Diaghilev wove into his multifaceted performances included composers such as Igor Stravinsky and Claude Debussy; artists such as Pablo Picasso, Joan Miro, and Henri Matisse; designer Coco Chanel; dancers and choreographers like Vaslav Nijinsky and George Balanchine, among many others.

Sgourev's (2015) study, which ultimately explores how local action both intentionally and unintentionally influences broader, field-level phenomena, begins by identifying the network dynamics that account for Diaghilev's unique ability to mobilize new action with broad impact. The study reveals that Diaghilev employed not only the *tertius iungens* and *tertius gaudens* brokerage models but also, as I will point out, conduit brokerage and knowledge articulation to get new things done. Using the BKAP model as a lens, I will briefly trace how this was done.

First, Diaghilev actively engaged in conduit brokerage by locating interesting and novel information and ferrying it back to Ballets Russes and its artistic productions. Quoting Beaumont (1940), Sgourev indicates,

> Diaghilev perpetually visited exhibitions and concerts, hurried to a private recital or to the studio of a still obscure painter, always listening and peering for the first indication of something new and important, which could be adapted to his productions and afford them a new interest, a new surprise. (2015, 350)

Recall in Chapter 1 that such a conduit-based search might involve the generation of entirely new ideas (i.e., for a new production) or, as in this case, be motivated and guided by an awareness of and desire to further a preexisting Ballets Russes production.

Alongside this curation of network-based artistic knowledge was the necessary melding of artistic ideas in any given performance. About Diaghilev's production of "Afternoon of a Faun," Sgourev (paraphrased from Munro 1951) writes:

> "The Afternoon of a Faun" was a complex work of art, a network of influences that spanned different historical periods and artistic forms. It featured late 19th century elements of Symbolism and Impressionism (Mallarmé, Manet), gestures borrowed from ancient Greek vases and Egyptian frescoes as well as from African tribal art, bright Slavic colors in Bakst-designed scenery and costumes, and stylized pacing and pauses emphasizing key moments in the performance. (2015, 351)

As the passage above suggests, tightly linked to this conduit brokerage was Diaghilev's "organizing acumen" (Sgourev 2015), which involved an active *tertius iungens* uniting of different artistic voices, as well as the artists, composers, and choreographers who stood behind them. As *tertius iungens*, Diaghilev united different aspects of a relatively fragmented avant garde artistic community, while also linking that community with the far more conservative but moneyed society elites who underwrote the artistic performances.

Finally, as a broker, Diaghilev did not unite all artists all of the time but selectively connected some and not others, a feat of orchestration and timing that involved a simultaneous measure of *tertius gaudens*. As Sgourev explains:

> In dance, painting and music, [Diaghilev] plucked his stars from a corps of novices and as soon as they had tasted fame, he kept the upper hand by playing them off against each other in competition for roles, money and visibility (Garafola 1989). Collaborators were valued as long as they had something new to offer—as soon as they ceased doing so, he felt no regret in parting with them (Grigoriev 1953, 64). (2015, 349)

To a great extent, therefore, Diaghilev's success was a function of his skillful blending of the three brokerage strategies in different combinations over time.

An analysis of Sgourev's data also reveals the role of knowledge articulation in Diaghilev's artistic process through the synthesis and shaping of a vast array of artistic content and voices. Consider four of the five communicative dimensions of knowledge articulation, namely, back stage/front stage; complex/simple; past, present, future; and familiarization/defamiliarization, as reflected in Sgourev's characterization of Ballets Russes performances:

> The sumptuous ballets choreographed by Fokine reassured high society in its tastes, while at the same time those choreographed by Nijinsky unsettled the public and critics, moving the boundaries of good taste (Huesca 2001). Even in the most adventurous productions, classicism and Modernism co-existed ... embedding Modernist experiments in choreography or music in a classical foundation, meshing provocations and tradition, bohemian artness and society chic (Davis 2006; Fermigier 1967). A key reason for the success of the Ballets in Paris and of Modernism, more generally, is that radical novelty was translated into a language rich in classical references, mixing and reconciling discourses, such as orientalism, symbolism, aestheticism, classicism and consumerism, accustoming the social elites to the meaning of artistic experiments and artists to the benefits of a rich and influential public. (2015, 353)

It is easy to infer the back-stage/front-stage dynamic of knowledge articulation, quite literally, in Diaghilev's strategic choices about which content to include in Ballets Russes performances and which to withhold. He made conscious choices to pull forward past influences (e.g., classicism) and merge them with future-oriented artistic experimentation, thereby generating creative tension. The knowledge translation and transformation that such a process required is also apparent in Sgourev's reference to "radical novelty [that] was translated into a language rich in classical references" (2015, 353).

The complex/simple dimension is also in evidence in the Ballets Russes' need to reconcile and simplify vast amounts of cultural knowledge while delivering complex syntheses that appealed to artists, elite theatergoers, and underwriters alike. Most salient, however, is the familiarization/defamiliarization dimension of knowledge articulation, with the familiarization accomplished by the company's emphasis on classical references that "reassured high society," strategically counterbalanced by defamiliarizing provocations of "bohemian artness." The tension inherent in "meshing provocations and tradition, bohemian artness and society chic" ultimately involves knowledge articulation along the familiarization/defamiliarization dimension.

Sgourev depicts Diaghilev's role as a mediator of supply and demand: a definition that anticipates the application of my model to early-stage entrepreneurship, to which we will turn later in this chapter. According to Sgourev, Diaghilev brokered between actors from the "supply side"—painters, dancers, and composers who might not otherwise know one another—while also connecting that artistic community with a "demand side": the high society that "consumed" the Ballets Russes "product" (2015). This more structural framing is consistent with the entrepreneurship literature's treatment of the broker as middleman, extracting profit from inefficiencies in the market (e.g., Casson 2010; Bonacich 1973; Ruef 2009[3]). That structural framing appears to continually coexist with the dynamic network process suggested by the Ballets Russes' dinner parties:

> The dinners after the performances brought together an "amazing worldly hodgepodge" (Nectoux 1992) of socialites, diplomats, industrialists, financiers, dancers, decorators, composers and writers that offered financial stability and cultural cachet to the Ballets (Davis 2006). (Sgourev 2015, 353)

Within Sgourev's analysis the creative project is readily apparent. The initial formation of the Ballets Russes, as well as its various productions, constitutes the creative projects that motivated the brokerage and knowledge articulation described above.[4] We can also intuit that over time new productions evolved from a "pure" raw creative project to a more routine-based activity and that certain meta-routines of coordination and production emerged within the company.

Diaghilev's Ballets Russes offers a compelling example of brokerage structure and process. Sgourev raises questions similar to those posed at the outset of this book regarding how local innovation scales and broad impact can emerge from local creative processes. Sgourev (2015) goes beyond the skilled broker as an agent of knowledge transfer (conduit brokerage), exploiter of differences (*tertius gaudens*), or integrator of local people, ideas, and resources in pursuit of innovative outcomes (*tertius iungens*). He points to the *tertius iungens* as one key to resolving the theoretical puzzle regarding how local, more microlevel action leads to more macrolevel outcomes involving broader field changes, collective benefits, or public goods (Cattani and Ferriani 2008; Stovel and Shaw 2012)—outcomes that might even outstrip the initial vision.

On this question of intent, Sgourev points out that Diaghilev's field-level impact on the arts could not have been predicted by his skilled—even inspired—but ultimately locally situated action.[5] Sgourev points out that Diaghilev's *tertius iungens* behavior, which comprised both structure and process, captured a catalytic property (Collins-Dogrul 2012; Stovel and Shaw 2012) of the *tertius iungens* that implicates growth on both a structural and process level. In exploring how local intent bridges to a broader impact, Sgourev references my observation (Obstfeld 2005) that the *iungens* broker, in introducing alters, forgoes the control of the *gaudens* while simultaneously exposing himself to potential alter-alter collusion (i.e., two alters, once introduced, can collaborate without or even against the broker who introduced them). Sgourev therefore suggests that the larger impact of Ballets Russes took place at a level at which Diaghilev had forfeited control.

A distinction worth noting here is that the *iungens* does not necessarily forfeit *all* control, but only a more immediate form of explicit, tangible control. Such looser control leverages broadened participation in the expanding, more dynamic network in which the broker can participate and intermittently facilitate but not strictly direct. It is this latter form of control, with its potential for faster, broader network growth, access, and impact, that the *iungens* broker accesses. Stated differently, the *iungens* trades the role of gatekeeper for that of impresario, ready to orchestrate multiple flows of collaboration, and the opportunity to foster but not control outcomes. Sgourev, citing Padgett and Powell (2012), uses the Ballets Russes case to suggest how *tertius iungens* brokerage may serve as a catalyst of systemic "tipping," or a chain reaction whose impact extends well beyond the broker's immediate efforts. This chain reaction phenomenon has obvious connections to considerations of how any case of mobilization, whether business related or collective-action related, might automatically scale.[6] While historical in nature, Ballets Russes' mobilization dynamics unfold in all "start-ups" as the BKAP's application to contemporary entrepreneurship research suggests.

Entrepreneurship Research Implications

The entrepreneurship literature currently presents many alternative conceptualizations of critical facets of entrepreneurship. Dynamic discussions, for example, explore alternative conceptualizations of opportunity (Shane and Venkataraman 2000; Shane 2012; Alvarez and Barney 2007; Alvarez, Barney, and Anderson 2013) and entrepreneurial action such as bricolage (Baker and Nelson 2005), effectuation (Sarasvathy 2008), and improvisation (Baker, Miner, and Eesley 2003). Despite the presence of these alternative conceptualizations, the entrepreneurship literature has yet to put forth an action-based social network perspective of entrepreneurship, a framework that accommodates and integrates these intersecting perspectives. The BKAP model can be employed to provide an integrated account of entrepreneurial action as consisting of microsocial action based on networks, brokerage process (in particular, *tertius iungens* processes of combination and recombination), and knowledge articulation, with the emergent firm conceptualized as a creative project. I will now briefly explore this as a jumping-off point for future research.

Over the past decade, a number of perspectives on entrepreneurial action have been put forth. One body of work has emphasized entrepreneurship as embedded in social actions (McMullen and Shepherd 2006), exploring how interactions among entrepreneurial actors during early stages may shape the direction of the entrepreneurial process (e.g., Gartner, Bird, and Starr 1992; Baker et al. 2003; Baker and Nelson 2005). These more recently emerging entrepreneurship perspectives present alternatives to prior approaches that emphasized the creation of new firms or wealth as a critical outcome of entrepreneurship (Alvarez et al. 2013). Ruef (2010) critiques another earlier approach to entrepreneurship that overemphasizes individual action while omitting the structural context of economic actors. As noted in the Introduction, as an alternative Ruef (2010) proposes an emphasis on the entrepreneur's intention to form a social group. This approach emphasizes the intentionality of entrepreneurs, as opposed to a discrete outcome or result associated with firm founding. Extending Gartner's (1989) emphasis on the emerging organization, Ruef argues for a process-based conception of how entrepreneurial groups emerge and operate, with a distinguishing focus on group-level behavior and collective action.

The role of combination and recombination is a well-established point of departure for entrepreneurship, and it is of particular use for expanding the action-based approach. According to Schumpeter, "Everyone is an entrepreneur when he actually carries out new combinations, and loses that character as soon as he has built up his business, when he settles down to running it as other people run their businesses" (1934, 78). In addition to underscoring the tension between the novelty of entrepreneurial undertakings as reflected in the creative project and the more routine-based work into which an entrepreneurial start-up may evolve as it stabilizes, Schumpeter's observation identifies a core function of entrepreneurship: the carrying out of new combinations.

To capture a microsocial view of brokerage activity, I return to Ruef's (2010) illustration of group formation, briefly noted in the Introduction, in a little more detail. Ruef sees such early-stage formation as the root of new venture creation. In Ruef's account, an entrepreneur, Luis Hernandez, persuades two collaborators, Bill Shipley and Diego Ramon, to begin a wholesale business for women's and children's clothing in Los Angeles. Luis's initial formation of this ownership core, along with his engagement of an investor and a supportive spouse, constitute the initial crucible of action characteristic of many start-ups. Though Ruef employs this example to illustrate some structural features of group formation, this example also illustrates critical, often-overlooked (and hence "invisible") aspects of the start-up social process: the triadic social network mechanisms by which Luis brings the initial entrepreneurial group together and the manner in which initial formulation and articulation of the potential start-up idea motivated that action. In this case, a central but often overlooked feature of the start-up is the connecting work that Hernandez pursues among founders, investors, collaborators, and other sources of critical resources that are necessary to the firm as it grows.

The Assembly View of Entrepreneurship

There are, I argue, two complementary forms of entrepreneurial combination at different levels of analysis: meta- and microsocial combination. At a meta-level, Schumpeter (1934) identifies five forms of combination that provide an overall logic for the formation of a new firm or market (Swedberg 2007).[7]

A combination might also refer to a more granular, early-stage microsocial formation of an entrepreneurial venture, such as the one described above. It is this second, more social sense of combination involving the unfolding activity of the entrepreneurial start-up that I refer to as an "assembly" view of entrepreneurship. Along these lines, I consider entrepreneurship to involve action shared by multiple individuals and rule out self-contained, single actor start-ups and the new firms they create. I also rule out the later, larger, and more stable stage of an emergent venture with established routines and greater certainty about its operating core. While such organizations might still be characterized as entrepreneurial along one or more dimensions, I focus on the salient differences between the start-up and a firm with stable, ongoing operation.

To get new things done in organizations, one must continually join both people and resources in new combinations (Casson 1982; Alvarez and Busenitz 2001; Shane 2012; Simmel and Wolf 1950). New organizing is an inherently *collective* process of joining, in an original way, the necessary people and the resources they bear. The *tertius iungens* orientation suggests the fundamental triadic mechanism associated with such brokerage activity. Entrepreneurial action involves sequences of *tertius iungens* combination and knowledge articulation in sequences of combination over time. In this assembly view, an entrepreneurial effort may "start" with the articulation of a new venture goal (or trajectory projection) but will unfold through subsequent combinations, over time, in the form of the creative project.

An entrepreneurial start-up necessarily begins with a trajectory projection, followed by cascades of *tertius iungens* connecting work in sequences of brokerage activity over time. This entrepreneurial action may evolve serially (i.e., with one combination leading to one or more subsequent combinations), in parallel (with new combinations developing simultaneously on multiple fronts), and expansively (with more people or resources being added to an expanding entrepreneurial core). In this latter sense, brokerage activity serves as the means by which new venture growth is accomplished.

Several additional points addressed previously in this book, including the *tertius iungens* linking that is crucial to early-stage entrepreneurial growth, are worth reiterating here. As with the creative project, an entrepreneurial trajectory will involve numerous brokerage episodes: some successfully realized and others that fail for any of numerous reasons. The combinations associated with the entrepreneurial start-up, as with the creative project more generally, are characterized by novelty and higher uncertainty than are those associated with the routine-based action that characterize the stable organization. Under these circumstances, knowledge articulation, always inherent to organizing, becomes critical as a means for coordinating and integrating parties in the absence of well-established routines. Finally, as with the network issues addressed in Chapter 1, *tertius iungens* linking may take place in either open or dense brokerage networks. It is for this reason that an assembly view of entrepreneurship may involve entirely new combinations (i.e., open networks) or, per Schumpeter (1934), recombination of people and resources in new combinations (i.e., dense networks).

The insight offered by this assembly view can also tie to two predominant entrepreneurship perspectives: bricolage and effectuation. Bricolage (Baker and Nelson 2005; Garud and Karnøe 2003) provides a resource-focused view of entrepreneurship in the face of scarcity. Baker and Nelson define bricolage as "making do by applying combinations of the resources at hand to new problems and opportunities" (2005, 333). The importance that the bricolage perspective has assumed in the entrepreneurship literature in part reflects its compelling depiction of the important entrepreneurial process of resource combination. I argue, however, that the bricolage process can be greatly clarified by revisiting resource combination and the associated combination mechanisms from a social network theoretical perspective, focused on actors connecting other actors and their associated resources.

Such an approach yields new opportunities to think more precisely about different resource combinations or combination sequences. It also suggests ways to differentiate between the network size, network density (or structural holes), and relative heterogeneity of trajectory elements (people, ideas, and resources) over time, distinctions that are relatively undeveloped in the bricolage literature. In light of bricolage's emphasis on resource constraint, the model presented here, by disaggregating actors and resource availability, allows a more precise unpacking of combinatorial action, resource availability, and the distinctions between processes associated with disruptive versus incremental innovation.

Effectuation (Sarasvathy 2008) is an entrepreneurial theory that shifts attention from the goals or ends of entrepreneurial action to the means available to the entrepreneur. To this point, Sarasvathy characterizes entrepreneurial action as "non-teleological—i.e., not taking preferences and goals as pre-existent or unchangeable," instead emphasizing "starting with the means and creating new effects" (73–74). Given this emphasis, Sarasvathy (2008, 2001) offers a strong critique of certain rational action theories involving "causal reasoning" (i.e., "beginning with a specific goal and a given set of means for reaching it") and prediction (assuming probabilities are "given or immutable").

Effectuation argues, in effect, for shifting attention to a view of action in which evolving means have the capacity to shape action more than do a priori goals or predictive calculation. In contrast to effectuation, the assembly theory introduced above views the trajectory projection as a critical forward stake in the ensuing creative project action trajectory that motivates and guides action. In this sense, I hold that the trajectory projection plays a crucial function in the entrepreneur's initiation and pursuit of action, though consistent with effectuation theory such goals are continually revised as action unfolds.

Future research could help differentiate between various types of creative project trajectories associated with emergent entrepreneurial action. Such trajectories, distinguished by the sequencing of people, resources, and idea combinations, may help to explain success and failure, rapid or slow growth, or the entrepreneurial patterns of action that match the complexity of their environments to a greater or lesser extent. Just as the BKAP model can help to tie seemingly disparate threads in the realm of entrepreneurship, so too can it serve as a unifying lens in another area of research revolving around mobilization: collective action.

Collective Action Research Implications

Gamson, Fireman, and Rytina (1982) describe an episode in 1936 that took place on the assembly line in General Motors' Flint, Michigan, Fisher Body plant number 1.[8] Gamson et al. recount Kraus's (1947) description of the heretofore unorganized auto workers' incipient collective action:

> The whistle blew [to start the shift]. Every man in the department stood at his station, a deep, significant tenseness in him. The foreman pushed the button, and the skeleton [automotive] bodies, already partly assembled when they got to this point, began to rumble forward. But no one lifted a hand. All eyes were turned to [the superintendent] who stood out in the aisle by himself. (1982, 2)

Kraus further recounts, "[The superintendent] seemed to become acutely conscious of the long line of silent men and felt the threat of their potential strength. They had been transformed into something he had never known before and over which he no longer had any command" (1947, 50). The incident, of course, did not happen in isolation but was part of an action trajectory that "began" with the firing of two welders the previous day and that within months escalated to the great GM sit-down strike of 1936–37, widely described as one of the most important labor conflicts in American history.[9]

As with the ethnographic cases presented earlier in this book, such a trajectory depicts joint action to get new things done. But whereas my ethnographic data depicted mobilization of white-collar engineers within an organization in order to disrupt and advance automotive design, here blue-collar workers mobilized action that ultimately disrupted and restructured an entire social and economic system. The vignette captures only one episode in a microsocial trajectory of action, comprising various acts of brokerage and articulation, which ultimately aggregated to the larger-scale collective action that followed.[10] As with Sgourev's study of Diaghilev's Ballets Russes, we could ask about the degree to which the sit-down strike was a function of the activists' intentionality throughout the process. We also would want to explore the extent to which mobilizing action was highly routinized, based on previous action, or unfolded in an improvisational middle ground where certain action repertoires (Tilly 1995) were deployed within a longer trajectory of action whose next steps were generally unknown and unfolded unpredictably. Though the BKAP model cannot account for all of the antecedents and outcomes of such a collective action trajectory, it does provide a theoretical and empirical framework to evaluate the more volitional, often nonroutine local action that is responsible for catalyzing such events and that can account for how they amplify (or fail to amplify) into larger outcomes.

The General Motors example is also noteworthy because it represents an example of the intersection of collective action and organizing, a nexus that is the subject of considerable recent attention among sociologists and organization theorists (e.g., Davis et al. 2005). Davis and Zald (2005) note that the underlying mechanisms of collective action are similar for both organizations and social movements. Put differently, there is much to suggest that innovation is not as "far removed from the street" (Clemens 2005, 362) as might be imagined. In the instance of getting new things done at the microsocial level, the mobilizing of action, whether in pursuit of innovation or collective action (or both), shares similar properties and processes. These properties and processes can be identified by means of the BKAP model, emphasizing brokerage and knowledge articulation as antecedents and projects or routines as outcomes. Not surprisingly, a model built to account for emergent innovative action applies to collective action that involves mobilization to drive social change.

First, both organizational innovation and collective action are motivated by the identification and removal of *grievances*, broadly understood. A grievance may concern a social injustice or a suboptimal product or process. Turner and Killian observe, "A movement is inconceivable apart from a vital sense that some established practice or mode of thought is wrong and ought to be replaced" (1972, 259). Organizational innovation and entrepreneurship, like collective action, often exhibit this "vital sense," though not necessarily with the dramatic contention often characteristic of collective action. Organizational innovation and entrepreneurship begin with the questioning of a product or process that has been taken for granted, and lead to engaging in new forms of coordination that "require ways of thinking and acting that are 'undoable' or 'unthinkable'" when first considered (Dougherty and Heller 1994, 202).

Similarly, the collective action literature's emphasis on the excluded status of challengers can be traced to the *mobilizing imperative* that motivates much innovation-oriented coalition formation in complex organizing and entrepreneurial ventures. McAdam's statement that social movements involve "rational attempts by excluded groups to mobilize sufficient political leverage to advance collective interests through noninstitutionalized means" (1982, 37) also can be said of much organizational innovation and entrepreneurship, where the excluded status of the challenger (Van de Ven and Hargrave 2004) drives those who lack "routine access to decisions that affect them" (Gamson 1975, 140). Such collective action is most suggestive of "bottom-up" or lateral mobilization innovation efforts, in which no single actor or function controls the focal product or process that the actors are attempting to change.

Though the absence of authoritative coordination is posed as one criterion for collective action (Clemens 2005), organizational theorists recognize that formal authority is often insufficient to drive many initiatives, necessitating coordination that involves more "top-down" or lateral mobilization by middle or upper managers (Strang and Jung 2005).[11] With respect to collective action, Jasper (2004) suggests that even those with considerable resources need to mobilize in a similar fashion to those with few resources. Smith (1996, 130–131), as cited by Jasper, describes the mixed challenger/insider status of many social movements:

> In fact, the majority of modern social movements possess moderate amounts of political and economic resources, enjoy limited access to political decision making, employ both disruptive and institutionalized means of political influence . . . and vocalize a mix of conciliatory, persuasive, and confrontational rhetoric. (Jasper 1997, 35)

Such a "mixed" status often characterizes innovation advocacy at many organizational levels. A larger takeaway from the commonality of innovation and collective action phenomena is the potential relevance of the action model put forth here to collective action theory, a relevance that has become more apparent due to recent developments in the study of contentious politics emphasizing collective action's underlying mechanisms and processes (McAdam, Tarrow, and Tilly 2001; Tarrow and Tilly 2007).

Contentious Politics, Collective Action, and Social Mechanisms

Over the past two decades, social movement theory converged around three factors driving social movement emergence: political opportunity structures, framing processes, and resource mobilization (McAdam, McCarthy, and Zald 1996).[12] A recent stream of work in the social movement literature has sought a systematic, microfoundational account of the many heterogeneous cases of social contention. Tarrow, for example, expressed concern regarding the social movement literature's "increasingly indeterminate agentic categories" and models that leave "pathways from structure to action vague and underspecified" (2003, 135). McAdam urged that social movement researchers make "a more serious investment in ethnography . . . and other methods designed to shed empirical light on the meso-level dynamics that largely shape and sustain collective action over time" (2003, 133). McAdam, McCarthy, and Zald indicate, "We have focused the lion's share of our research energies on the before and after of collective action. . . . But we haven't devoted a lot of attention to the ongoing accomplishment of collective action" (1988, 729). In a similar vein, Snow (2006) argues for the need to better specify how different aspects of social movements (e.g., political opportunity, mobilizing structures, and framing) dynamically interact.

In response to many of these types of concerns, McAdam et al. (2001) published an ambitious work that identified fundamental mechanisms and processes that could be used to establish similarities in cause-effect relationships that underpin a wide variety of contentious politics cases. The authors' reference to mechanisms drew on Hedström and Swedberg's (1998) call for a more systematic pursuit of the social mechanisms that generate and explain observed associations between events.[13] McAdam et al. (2001) suggested a number of mechanisms that underpinned collective action and contentious politics, among which brokerage figured quite prominently. They located these mechanisms within "processes" the authors described as "sequences and combinations of causal mechanisms" (McAdam et al. 2001, 12–13).

There is a rough correspondence between the idea of processes comprising sequences and combinations of mechanisms, on the one hand, and my conceptualization of action trajectories comprising combinations of elements (e.g., people, ideas, resources, and artifacts), on the other, a similarity to which I will return below. Tilly and Tarrow (2007) revisited and sharpened this ambitious approach in a second book that emphasized three basic mechanisms—brokerage, diffusion, and coordinated action—that align with my BKAP action model. Tilly and Tarrow define the first of their key mechanisms, "brokerage," as the "production of a new connection between previously unconnected sites" (2007, 31), a definition consistent with the *tertius iungens* orientation. The authors' second mechanism, diffusion, defined as the "spread of a form of contention, an issue, or a way of framing it from one site to another" (31), corresponds with some combination of what I call conduit brokerage and knowledge articulation. Tilly and Tarrow do not address the *tertius gaudens* brokerage, but I assert that collective action cannot be completely understood (or analyzed) without tracing the intertwining of *iungens*, *gaudens*, and conduit brokerage strategies alongside knowledge articulation.

By emphasizing broker behavior, my treatment of collective action is more microsocial and agentic than that of Tilly and Tarrow, who refer more broadly to brokerage and diffusion as occurring between "sites," an approach that may be more conducive to capturing more macrolevel action trajectories. The BKAP model's separate capture of social network structure and process offers greater analytical precision with respect to evaluating collective action emergence. Such a process accommodates a long tradition of social network analysis in collective action and social movement literatures, while offering the opportunity here to explore the interaction of both structure and process longitudinally.

Tilly and Tarrow refer to a third major mechanism, coordinated action, which the BKAP model also addresses, at least partially. The authors define coordinated action as "two or more actors' engagement in mutual signaling and parallel making of claims on the same object" (2007, 31). It is interesting to note that for this third mechanism the authors switch from "sites" to "actors" as the analytic object. This more micro facet of coordinated action is addressed by the double and triadic interact dynamics that I explored earlier, but might easily scale to the larger social phenomena (see treatment of catalysis and autocatalysis later in this chapter).

My approach to brokerage and knowledge articulation, along with my distinction between projects and routines, also suggest an opportunity to situate social movement theory's concept of action repertoires within a creative project trajectory. Tilly defines action repertoires as "the established ways in which pairs of actors make and receive claims bearing on each other's interests" (1995, 43). This approach refers to routines that individuals or groups actively construct to accomplish their objectives. Despite more routine ways of coordinating, anticipating, representing, and interpreting each other's actions, however, collective action repertoires also involve continual improvisation in tactics, messages, and self-presentation (McAdam et al. 2001). The characterization of an action repertoire as routine, homogeneous, and stable may reflect what is salient from a more macro or historical perspective. Either way, action repertoires can be seen as first-order routines that are part of a larger meta-trajectory of emergent collective action.[14]

In the aforementioned ways, the BKAP model may be employed to further theorize and analyze collective action phenomena in terms of brokerage and knowledge process. Such a perspective expands our understanding of collective action brokerage beyond what Oliver suggested was the "usual definition of brokerage," that is, "people who help parties with partially conflicting interests find a mutually agreeable bargain" (2003, 122). I follow this line of inquiry regarding how the BKAP model illuminates different forms of mobilization into one final arena to address a theme noted in the Introduction: the relationship of traditional and technologically transformed forms of networking.

Mobilizing Action in an Analogue-Digital World: The Case of the Arab Spring

As noted in the Introduction, digital tools have clearly transformed and will continue to transform organizing in ways that are difficult to fully grasp or anticipate. With the dramatic increase in the availability of knowledge made possible by digital infrastructures, the broker's role sometimes shifts from leveraging scarce knowledge to curating large volumes of knowledge. Moreover, digital infrastructures sometimes shift the work of coordination from the shoulders of *brokers* to that of digital *venues* (e.g., Facebook, LinkedIn, Twitter, and a myriad of smaller niche platforms) that identify people with common interests, shared histories, or mutual ties, without hands-on brokered introductions. Common interests may reflect foci (Feld 1981, 1982) such as offices or voluntary organizations, or settings (Sorenson and Stuart 2001) such as research consortia, trade associations, and boards of directors that give rise to interorganizational ties. Of course, venues did not emerge in the digital age—marketplaces and town squares have been around for millennia—but the role of digital venues in re-shaping organizing is difficult to overstate. In such digital venues, rapidly evolving technology may accelerate introductions through sophisticated algorithms that match users or assess reputation through user feedback.

The proliferation of such digital venues, however, does not eliminate the advantages of "hands-on" brokering within those venues or of orchestrating projects across multiple venues. Also, *creating* new digital venues that ultimately facilitate hundreds, thousands, or millions of interactions may often still require extensive in-person brokerage. The entrepreneur/broker/strategic actor of the future, it appears, will navigate both the in-person and internet worlds, what I sometimes loosely refer to as *mobilizing action in an analogue-digital world*. By analogue skills, I refer to the social skill (i.e., brokerage and knowledge articulation) detailed in this book. By digital skills, I refer to the ability to work skillfully on such platforms as Facebook, LinkedIn, and Twitter, with or without a complement of programming skill.

Let's look at a sustained case in point. Egyptian activist and Google executive Wael Ghonim's role in the Arab Spring provides a rough feel for what such analogue-digital straddling entails and how newly emerging forms of digitally infused collective action might unfold. In traditional terms of social movement theory, the 2010 Egypt that Ghonim (2012) engaged (from his location in Dubai) displayed a political opportunity structure ripe for contention.[15] Yet while the structural antecedents were present, 2010 Egypt displayed neither the resonant movement frames nor the mobilized resources necessary for broad collective action to emerge.

As an actor outfitted with both analogue and digital skill, Ghonim was well positioned for his activist role. Born in Egypt and married to an American, and as Google's top marketing representative for the Middle East, he was "multi-lingual" (in the broadest sense of the word) in a network with many structural holes. Simultaneously occupying Middle Eastern and Western cultures, Ghonim commanded a portfolio of digital skills that included programming but also, nontrivially, the ins and outs of texting, Facebook, and operating on the internet more broadly.

In the face of rising social discontent, Ghonim skillfully created, in sequence, two high-impact Facebook pages, recognizing that Facebook pages were a more effective device than Facebook groups.[16] His first "digital" move was the creation of a Facebook page supporting Nobel Peace Prize-winning nuclear weapons expert Mohammed El Baradei, a symbol of Western talent and credentials and a critic of the Mubarak regime. The stage for the "El Baradei"

Facebook page was inspired by a prior face-to-face ("analogue") meeting between Ghonim and Baradei and was later followed by the creation of a closed
Google e-mail group ("ElBaradei") to facilitate confidential communication
among supporters.

Ghonim next created a Facebook memorial page for Khaled Said, a young
Egyptian man who died while under police custody, entitled "We are all Khaled
Said." This page attracted hundreds of thousands of followers, with many Egyp-
tians using his photograph for their own Facebook profiles. The digital success
of Ghonim's page somewhat obscures the invisible analogue work that enabled
it. The naming of the website itself, along with the language of protest used on
it, started with well-cultivated "analogue" social skill (per Chapter 6), including taking the attitude of the other (Mead 1934), seeing the "self" (in this case,
the yet-to-be posted web page) from the eyes of the other (i.e., Egyptian web
page viewers), conducting thought experiments about how various Facebook
page features would be received, and making appropriate adjustments in page
design. Ghonim's instincts, for example, told him that first-person language
would work best:

> I spoke on the page in the first person, posing as Khaled Said. What drove me,
> more than anything else, was the thought that I could speak for him, and if even
> a single victim of the regime could have the chance to defend himself, it would
> be a turning point. Speaking as Khaled gave me a liberty that I did not have on
> ElBaradei's quasi-official page. It also had greater impact on the page's members.
> It was as though Khaled Said was speaking from his grave. (2012, 60)

Ghonim's view of the self as other, along with his many-faceted multilin-
gualism, provided other crucial insights about how to create the Facebook page
before him:

> Even though I was proficient at classical Arabic (al-fusHa) from my school years
> in Saudi Arabia, I chose to write my posts on "Kullena Khaled Said" in the col-
> loquial Egyptian dialect that is closer to the hearts of young Egyptians. For the
> generation born in the eighties and nineties, classical Arabic is a language read
> in the newspapers or heard during news reports on television and comes across
> as quite formal. By using colloquial Egyptian, I aimed to overcome any barriers
> between supporters of the cause and myself. I also deliberately avoided expres-
> sions that were not commonly used by the average Egyptian or that were regu-
> larly used by activists, like *nizaam*, the Arabic word for "regime." I was keen to
> convey to page members the sense that I was one of them, that I was not differ-

ent in any way. Using the pronoun *I* was critical to establishing the fact that the page was not managed by an organization, political party, or movement of any kind. On the contrary, the writer was an ordinary Egyptian devastated by the brutality inflicted on Khaled Said and motivated to seek justice. This informality contributed to the page's popularity and people's acceptance of its posts. (61)

This second page's success was clearly a function of analogue-digital skill deployed with rapid cycles of experimentation and learning.[17] A subsequent stream of social innovations that straddled the analogue-digital divide reflected Ghonim's instincts for and translation between the two domains, including polling Facebook page users for ideas, publicizing rallies through printed fliers and mass texting, and posting times and locations for demonstrations hours before they occurred to stay a step ahead of the police. The viral response to the Khaled Said and Mohammed El Baradei Facebook pages may appear to the casual reader as inherently "self-organized" and entirely removed from any analogue efforts, but that would be a wholly inadequate interpretation.

Ghonim's analogue-digital engagement can be described in terms of the BKAP model's variables. His network spanning a Western corporation (Google) and Middle Eastern society featured an abundance of structural holes that co-existed with dense clusters of trust and knowledge exchange. His analogue-digital activism displayed a powerful composite of brokerage skill that involved moving knowledge (conduit), mobilizing action both in person and digitally (*tertius iungens*), and crucially, continually keeping authorities separate, in the dark, as a critical ongoing strategic imperative (*tertius gaudens*). His knowledge spanned the analogue-digital divide, and as detailed above, he was masterful in using analogue instincts to craft highly impactful digital messages. The collective action Ghonim helped shape can be understood as a form of a creative project, often developed in the relative absence of established collective action routines, to mobilize support in a rapidly evolving context.

Over time, certain more routine approaches no doubt emerged (i.e., the second and third time that mass texting was employed to generate a demonstration), but there were no such action repertoires (Tilly 1995) at the outset. In addition, the goal of the insurrection, the trajectory projection itself, was continually evolving, as what was necessary and possible evolved along with the identities of the activists themselves (i.e., evolving means and ends). Such a highly uncertain project trajectory likely unfolded in a fashion similar to the way in which contentious politics evolved in the 1936 General Motors sit-down strikes (discussed earlier) or, for that matter, many new Silicon Valley start-ups.[18]

Research Implications for Addressing
Social Inequality through Education

Considering GM and the Arab Spring, you can see how the BKAP model has not only important theoretical implications but also practical and social ones. The model's emphasis on social skill logically extends to the performance of disadvantaged college students as well. Recent work has placed rising income inequality (e.g., Piketty 2014; Stiglitz 2012) at the forefront of public awareness. Though this work ultimately emphasizes macrostructural issues, it also raises related questions about how such inequality might be addressed.

A social network-based perspective on social skill may help, for example, particularly in the context of education. Higher education remains the primary institution through which income inequality is both minimized and legitimated (Bourdieu 1996; Massey 2005). The "intergenerational transmission of advantage" concerns the advantages associated with differential financial investments in children's enrichment, quality of schooling, and access to superior college options through which such a hierarchy is created and replicated. These advantages are further reinforced through the superior instruction and access to resources afforded students at elite universities. On the other hand, recent work has shown that a bachelor's degree continues to minimize the direct effects of socioeconomic origins on socioeconomic success in the United States (Pfeffer and Hertel 2014; Torche 2011). In other words, while students from disadvantaged backgrounds are less likely to attain a college education, those that do appear to enjoy substantial social mobility.

The BKAP model directs attention to social network processes, such as relationship formation, facilitation of network cooperation, and knowledge transfer, that work in conjunction with social network structure to influence performance and attainment. While social capital and network structure are important determinants of performance, it is equally important to grasp and augment the social processes through which disadvantaged students (e.g., first generation and underrepresented minorities) mobilize support and otherwise advance. Disadvantaged-student success will be a function not only of individual differences (e.g., intelligence and grit) and brokering-based social skill but also of the ability to enlist brokering support through which teachers, administrators, alumni, and institutions connect the student to resources, opportunities, social support, and identity development (Small 2009). Obviously, universities themselves present disparate advancement contexts through which access to resources and social skill development may or may not be provided.

Interestingly, the most important application of the BKAP model may be to students at comprehensive universities, in contrast to research (some might say "elite") universities. Comprehensive universities, that is, regional, public, four-year institutions with joint teaching and research missions, tend to be lower cost and to educate a mix of traditional and nontraditional students. Comprehensive university students display considerable resilience to first advance through community colleges to arrive at universities, and they represent a unique, highly motivated population poised to sustain dramatic upward mobility.[19] Such a population, and not the well-endowed student afforded the full benefits of the intergenerational transmission of advantage, represent the true "cream" of American society that social skill and brokering could further assist in substantial advancement. Comprehensive university students also represent a substantially larger population than the one tied to research universities. Formal study and programmatic assistance, through attention to social networks and social skill, suggest a particularly powerful lever for theorizing mobility and broad remediation of income inequality. Rigorous research designs could expand on the BKAP model to study the degree to which social network structure and social skill differ across different populations and settings and how they develop differentially over the early adult life cycle, from university into early employment. Such an expanded approach would examine how disadvantaged students' social skill, parental social capital, and university context influence educational attainment.

Concluding Thoughts

These applications—which take us from the arts to the business sector to the political arena and finally to education and social inequality—lead us back to the Introduction. Recall that I referenced Boltanski and Chiapello's (2005) compelling assertion that we live in an era increasingly defined by project-based work. The authors describe this as a projects-oriented era where organizing, and life itself, revolves around projects, networks, and reduced hierarchy, with a premium placed on the ability to function as a broker.

This perspective speaks to a widely intuited realization that technology-enabled organizing and living is very different from what it was even a decade ago, with increasingly open communication and coordination only accelerating. If we examine these observations with the BKAP model in mind, we can see that while the explanatory variables of network structure and individual knowledge have always been central to action, the social skill associated with orchestrating networks and articulating knowledge, along with the dependent variable concerned with formulating and pursuing novel projects, are of increasing importance in these newly emerging and rapidly evolving contexts.

What is needed is a model of action that captures the fundamental relationship between structure and process in the emergence of innovative action across multiple contexts, a model that is sufficiently specified but still flexible enough to have broad application. The parsimonious BKAP framework identifies the creative project as the form that many kinds of newly emerging action take. The creative project provides a construct with sufficient theoretical and empirical mettle to address a range of new action with novel, less routine origins, while simultaneously providing a conceptual tether to the stability associated with the organizational routine, a familiar construct at the center of organizational theory for more than half a century.

The BKAP model may strike some as simplistic. One might ask, for example, whether there is an interaction effect between networks and knowledge. My answer would be, "Of course, depending on the context," although I view this as a normal science puzzle that should take a temporary backseat to explicating the larger theoretical framework set forth here.

The quantitative measures briefly noted in Appendix 2 are as follows: (1) provide justification based on field observation and quantitative analysis; (2) lay the groundwork for subsequent quantitative and qualitative studies with and without interaction effects; and (3) support new measures that capture the full impact of the three brokerage action orientations alongside knowledge articulation. In this book, however, I have emphasized the articulation of a conceptual model that reveals different social phenomena from those that have been identified in the organizational research and neighboring fields to date.

For the BKAP model of local action to be relevant, one has to be persuaded first that it has broader generalizability to creative projects and routines in multiple contexts, and second that it identifies a nexus of action that catalyzes or scales to larger social phenomena. This latter question concerns intangibles such as political opportunity structure and identity with which the social movement literature has long grappled. A third puzzle concerns whether and when creative projects, when they do emerge, account for significant change. Emirbayer and Mische address this puzzle when they observe, "Actors who feel creative and deliberative while in the flow of unproblematic trajectories can often be highly reproductive of received contexts," and relatedly, "Actors who feel blocked in encountering problematic situations can actually be pioneers in exploring and reconstructing contexts of action" (1998, 108–109).

In a similar vein, Swidler (1986) observes that unsettled lives and periods of social transformation create a context for new strategies of action. Fligstein and McAdam (2012) contemplate a similar paradox that while social skill is always found in stable fields and in the creation of new ones, the most opportune time for entrepreneurial action is in emerging or destabilized fields. The key issue here, I would argue, is to become more theoretically and empirically precise about the nature of social skill but to pay equal attention to the distinction between entrepreneurial projects that replicate the existing social order versus those that reshape it.

As noted, the *tertius iungens* construct has a natural alignment with conceptualizations of "combination" as used by Schumpeter (1934) and "articulation work" (Strauss et al. 1985; Star and Strauss 1999; Suchman 1996). Both circumambulate the phenomenon of how "things," that is, people, resources, and ideas, are brought together in some alchemy of networks and knowledge. This volume constitutes an effort to move beyond alchemy to science, not only as science relates to routine processes of combination but also how it relates to more elusive processes of novel, emergent combination. The stakes are high. As de Tocqueville (1966) observed, "In democratic countries knowledge of how to combine is the mother of all other forms of knowledge." I take de Tocqueville's observation as advance support for this project, but also suggest that such knowledge is important not only in democracies but in all political and organizational contexts, presuming that the necessary constraints are taken into account.

It turns out that "knowledge of how to combine" is not the most recent and widely accepted translation of de Tocqueville's prescient observation. The phrase de Tocqueville used can be more accurately translated as a "science of association," based on Goldhammer's translation: "In democratic countries, the science of association is the fundamental science. Progress in all the other sciences depends on progress in this one" (de Tocqueville 2004, 517). Tocqueville goes on to say:

> When citizens can associate only in certain instances, they regard association as a rare and unusual undertaking, and it seldom occurs to them to consider it. When they are allowed to associate freely for any purpose, they ultimately come to see association as a universal and, as it were, incomparable means of achieving the various ends that mankind proposes for itself. Each time a new need arises, the idea of association comes immediately to mind. The art of association then becomes, as I said earlier, the fundamental science; everyone studies it and applies it. (2004, 522)

I believe that the importance of the "science of association" requires no such qualification. I would assert categorically that the art of association is the master social science. The substitution of "science" for the more amorphous "knowledge" used in the earlier translation of de Tocqueville puts things on the correct footing. It is through a systematic theoretical and empirical treatment, an approach I have endeavored to provide here, that we are in a better position to understand how people get things done, how they get new things done, and how much-larger processes such as collective action, entrepreneurial action, and democracy unfold. If the effort here can be construed as revealing new things about the science of association, I will deem it a success.

Appendix 1
Qualitative Methods

Research Setting and Data Collection

Field observations were conducted at NewCar, a division of the major automotive manufacturer AllCar with more than one thousand employees, with particular attention given to four hundred who were dedicated to the design of a new vehicle, the G5. (All persons' names, organizational names, and acronyms are pseudonyms.) I began collecting field observations midway through the company's approximately five-year automotive design process. Intensive field observation (four days a week) lasted nine months followed by four days a month for the next fifteen months. Approximately one year after beginning my fieldwork at NewCar, I conducted a large-sample survey study to test hypotheses generated from my field observations.

Overall, I spent approximately one thousand hours in observation, participation, and conversation. Data analyzed here derive from field notes; formal but unstructured interviews; and organizational documents. Each day of observation yielded five to fifty pages of handwritten field notes, which I usually wrote up within twenty-four to forty-eight hours after leaving the field. These field notes were supplemented with several hundred interviews, more than a hundred of which were taped and transcribed. As part of the data-gathering effort, I routinely collected documents such as memos, meeting minutes, prints of CAD/CAM designs, and informal sketches.

I was given almost unlimited access to divisional meetings and attended a wide variety of meetings in each of the seven major engineering groups. Regular meetings with the G5 program manager—my principal sponsor—served to orient me toward the structure and activities of the organization. My attendance at formal meetings was supplemented by observation of informal postmeeting conversations, multiple informal meetings, and periodic extended observation of individuals' work.

Data Analysis

Inductive analysis (Strauss and Corbin 1990; Locke 1996; Fox-Wolfgramm, Boal, and Hunt 1998; Dougherty 2002) of field observations throughout the observation period led me to identify themes and categories that were continually developed to reflect new, incoming data. My preliminary analysis identified several antecedents to innovation and change. Among those, coordination behaviors and innovation-directed communication were central to innovation advocacy. Field notes and interview transcriptions were coded according to these emergent categories, and further iteration between data and theory suggested labeling the key behaviors *tertius iungens* brokerage and *knowledge articulation* (Polanyi 1958; Winter 1987). Analysis of approximately fifty innovation advocacy episodes yielded the knowledge articulation dimensions described in Chapter 2. This data analysis served as the basis of constructing corresponding measures for a subsequent survey study (Obstfeld 2005) conducted in the same firm after most fieldwork was completed and analyzed on a preliminary basis (see Appendix 2). The results of that survey study both validated the knowledge articulation and brokerage observations, and further fed back into an ongoing inductive, theory-generating approach. This triangulation (Jick 1979), for example, led to the insights that (1) *tertius iungens* brokerage took place in both open and closed networks, and (2) *tertius iungens* and knowledge articulation were closely related.

Appendix 2
Quantitative Survey Insights

In previous research (Obstfeld 2005), I used survey data to predict the successful mobilization of NewCar innovation as a function of network process (i.e., *tertius iungens* brokerage) alongside network structure (i.e., density or effective constraint) and individual knowledge. I collected and analyzed these survey data after having completed most of the NewCar fieldwork, and the survey findings supported, along with the qualitative findings, the development of the BKAP framework explored in Chapters 1, 2, and 3.

In the survey study, I used ordered logit regressions (Obstfeld 2005) to predict engagement in innovation, measured as innovation involvement. The dependent variable, innovation involvement, measured respondents' self-reported, highest level of participation across seventy-three G5 product or process changes, using Ibarra's (1989, 1993) five-category scale of innovation involvement: (1) initiator of the innovation, that is, if its introduction and use were in large portion your idea; (2) major role in innovation; (3) minor role in bringing the innovation to the organization; (4) know of innovation but had nothing to do with it; and (5) the innovation is one you know nothing about. Respondents' self-report responses were then validated against a second survey that asked G5 engineers to evaluate who was the source of each innovation.

Results indicated that *innovation involvement* was significantly predicted by network structure (e.g., density), network process (*tertius iungens* brokerage), and knowledge (as measured by social knowledge, technical knowledge, years in the firm, and education) (see Table A.1). The *tertius iungens* action orientation was measured by a scale created in part from insights derived from my field observations. The *tertius iungens* scale (Obstfeld 2005) consists of the following six items: (1) I introduce people to each other who might have a common strategic work interest; (2) I try to describe an issue in a way that will appeal to a diverse set of interests; (3) I see opportunities for collaboration between

Table A.1.   Models Predicting Innovation Involvement

Ordered Logit Coefficients Predicting Innovation Involvement ($N = 152$)		
Variable	Model 1 (with Density)	Model 2 (with Constraint)
Social Network		
Density	1.954** (.850)	--
Number of alters	-0.005 (.043)	--
Constraint	--	2.079 (1.609)
Tertius iungens orientation	0.324* (0.183)	0.314* (0.182)
Knowledge		
Social knowledge	0.562*** (0.142)	0.525***
Technical knowledge	0.088 (0.121)	0.108 (0.119)
Years in firm	0.061* (0.026)	0.063** (0.026)
Years in firm (dummy variable)	0.930* (0.532)	0.939* (0.531)
Education	0.619* (0.371)	0.584 (0.371)
Organizational Rank	0.424* (0.215)	0.487* (0.213)
Chi square	71.701***	67.725***
D.f.	9	8
Nagelkerke R^2	.401	.383

*$p < .05$; ** $p < .01$; *** $p < .001$
Source: Unofficial figure from defunct automotive firm, 1988.

people; (4) I point out the common ground shared by people who have different perspectives on an issue; (5) I introduce two people when I think they might benefit from becoming acquainted; and (6) I forge connections between different people dealing with a particular issue.

That 2005 study, however, did not incorporate a measure for knowledge articulation elaborated in Chapter 2. Toward the end of my NewCar fieldwork, qualitative analysis of approximately fifty innovation advocacy episodes taken from field observations and interviews suggested the importance of knowledge articulation and the relevance of different "articulation devices" as the means by which knowledge articulation was accomplished. Before leaving NewCar,

I distributed a short follow-up survey, which 108 of the original 152 respondents completed. The main question asked of those previous respondents was, "To what extent in a one-hour meeting do you engage in the following?" The possible responses included, "analogies and metaphors," "stories," "PowerPoint," "2-D or 3-D software," "informal graphs or sketches," "humor," and "physical parts"—all behaviors that my field observations indicated were often found within successful knowledge articulation episodes.

Though as yet unpublished, findings from this analysis suggest the validity of a knowledge articulation measure. Factor analysis revealed that "analogies and metaphors" and "stories" load onto a discrete factor. Further analysis indicated a high correlation between this knowledge articulation measure and innovation involvement, consistent with the impact of knowledge articulation in my field observations. In addition, this knowledge articulation measure and the *tertius iungens* orientation were highly correlated. I interpret this relationship as suggesting that *tertius iungens* brokers often employ analogies, metaphors, and stories to facilitate interaction between alters. In summary, these unpublished analyses provide consistent evidence regarding how networks and network processes, alongside knowledge and knowledge processes, serve as interlocking aspects of social skill necessary to get new things done.

Notes

Introduction

1. Taleb (2007), for example, points out that while "black swan" events are comparatively rare occurrences, they have a disproportionate ability to reshape social systems.

2. Gould and Fernandez's (1989) foundational work unpacks the brokerage phenomenon through five fundamentally distinct brokerage types that are distinguished by the different memberships that the broker and each of the two alters might have (i.e., coordinator, itinerant broker, gatekeeper, representative, liaison). Recognizing that different parties to a brokerage arrangement hold different interests, Gould and Fernandez use subgroup memberships, whether shared or distinct, to distinguish between these brokerage categories. While it is beyond the scope of this volume to fully engage this important work, consider some of the strengths and limitations of the Gould and Fernandez brokerage scheme. First, the approach, by tracing subgroup memberships of all three parties to an open triad, provides simple, objective quantifiable distinctions to further distinguish between five variations on the open triad structure. Second, by formalizing these five variations on the open triad, the authors elucidate important implications of differing membership and alignments in these coordinative contexts. Third, the Gould and Fernandez model captures a complexity in the brokerage role that involves both the individual and the subgroups from which the different parties originate, speaking simultaneously to the microfoundations and interorganizational contexts of many brokerage cases.

There are, however, several limitations with the fidelity of the Gould and Fernandez model to brokerage phenomena that speak to the alternative empirical and theoretical approach provided in this book. First, the Gould and Fernandez model is based entirely on structural distinctions specifically with respect to the open triad (i.e., uncompleted two-paths); while their approach substantively advances brokerage analysis by allowing for the evaluation of different structurally-defined brokerage opportunities, it is agnostic to the specific brokerage processes involved (Spiro, Acton, and Butts 2013). Second, Gould and Fernandez's cross-sectional design does not allow for variation in the social process within those structures or their empirical consequences as brokerage process evolves over time. Third, Gould and Fernandez assume clearly defined membership roles when brokerage roles are, almost by definition, ambiguous, ill-defined, multiple, hidden, and evolving. Finally, Gould and Fernandez assume that brokerage, by definition, involves a non-tie between alters. Despite the importance of the non-tie condition, I argue that brokerage can also take place in the presence of the alter-alter tie, a condition that the predominant definitions of brokerage rule out.

3. *Tertius iungens* (YUNG-gains) is based on the Latin verb "*iungo*" which means to join, unite, or connect. In early Latin, it means literally to yoke, harness, or mate and serves as the root of such modern words as junction, conjugal, and yoga. In one context it is used in the phrase "to throw a bridge over a river." In later Latin, it seems to be used in a more metaphorical sense, "to unite" or "to form" (as in a friendship). Cicero used the phrase "*iungere amicitiam cum aliquot*," that is, "to form a friendship or alliance with another." For more on the origins of *tertius gaudens*, see Burt (1992). Both of these orientations will be discussed in great depth in the next chapter.

4. All persons' names, organizational names, and acronyms in this book are pseudonyms.

Chapter 1

This chapter is an extension of Obstfeld, Borgatti, and Davis (2014).

1. Where Parsons stresses the importance of the dyad, he attributes great importance to the potential for culture and socialization to provide "the grounds of consensus" (Parsons 1966, 14;Vanderstraeten 2002). Luhmann (1995), on the other hand, sees more agency for shaping the interaction within the dyad.

2. In his famous essay "The Triad," Simmel (Simmel and Wolff 1950) observes how the arrival of a child substantively alters the previously dyadic nature of the marriage.

3. This literature has emphasized that invisible work is often done by women.

4. My study of new product development in the automobile industry, described in the Introduction, found that automotive engineers had cohesive (dense) networks devoted to the design of a given subassembly, which allowed for the most efficient coordination and exchange of complex knowledge necessary to complete error-free designs.

5. I take as a point of departure Burt's (1992) influential introduction of structural holes theory, in part because it continues to have such a sustained impact on the social networks literature, and in particular the structural view of brokerage. I note, however, that Burt and others have continued to develop structural holes theory, both empirically and theoretically, in the two decades that have ensued. Because of the range and depth of this subsequent work (e.g., Burt 2005, 2010), I will not attempt to recapitulate these developments but to acknowledge its ongoing importance and, as well, a sophisticated process perspective that informs it. Later, I briefly note how the process perspective intersects with recent work by Burt (i.e., Burt, Merluzzi, and Burrows 2013).

6. I take the idea of the "token" as a generic term for an idea, a story, or a practice from Latour: "According to [the model of translation], the spread in time and space of anything—claims, orders, artifacts, goods—is in the hands of people; each of these people may act in many different ways, letting the token drop, or modifying it, or deflecting it, or betraying it, or adding to it, or appropriating it. . . .When no one is there to take up the statement or token then it simply drops" (1986, 267). Latour uses the concept of the token to problematize more conventional views of diffusion that hold ideas and objects as diffusing through society unchanged.

7. Carlile's framework emerged from an ethnographic study of within-firm product development across functional boundaries, but it is equally applicable to the cross-boundary challenges confronted by the broker operating between two or more alters in any context. Rogers raises similar issues, noting that diffusion is influenced by "the degree to which pairs of individuals who interact are similar in certain attributes, such as beliefs, education, social status, and the like" (1983, 18).

8. Multiple meanings have been imputed to Simmel's original terminology. I employ "*tertius gaudens*" here to refer to those cases where playing alters against one another is the broker's focus, and avoid using it to reference a broader leverage often commanded by a third party.

9. In this and subsequent illustrations, the boldfaced role characterizations are my own additions.

Chapter 2

1. This first meaning corresponds to the articulation work literature (Strauss et al. 1985; Gerson and Star 1986; Suchman 1996), which examines efforts to coordinate people and their associated work. The articulation work literature is closely linked to the idea of invisible work noted earlier, which emphasizes how crucial coordinative work is relegated to less powerful and often female workers—a somewhat paradoxical contrast with accounts of brokerage as a means by which strategic actors gain or wield influence, often to get new things done.

2. These two definitions of "articulate" are both from the *Oxford English Dictionary*.

3. The word "combination" also signals the close connection of brokerage and knowledge process. In many contexts, it is used to describe how actors or firms combine people in networks (e.g., Hargadon and Sutton 1997; Fleming, Mingo, and Chen 2007; Burt 2004; Obstfeld 2005), but in other contexts it addresses how firms or individuals synthesize and apply current and acquired knowledge (e.g., Kogut and Zander 1992; Rodan and Galunic 2004; Fleming and Sorensen 2004). Speaking to the failure to fully connect the study of network and knowledge processes, Mische and White point out that the two "approaches are ripe with mutual resonances and implications, yet they have so far maintained a skeptical aloofness from each other in regards to research strategy and design" (1998, 695).

4. From a network perspective, two out of four dimensions from Granovetter's (1973) classic operationalization of tie strength correspond with common measures: "amount of time" (that the tie has been in existence) and "emotional intensity." These two contrast with Granovetter's two other tie strength operationalizations, which emphasize the back-and-forth interaction (social process) between two nodes: "intimacy" (mutual confiding) and "reciprocal services."

5. These issues revolve around whether tacit knowing is completely or even partly articulable, whether tacit and explicit knowledge constitute separate and distinct forms of knowledge, and relatedly, the convertibility of knowledge between its tacit and explicit forms (Tsoukas 1996; Cook and Brown 1999; Orlikowski 2002). Polanyi (1958) devoted a considerable amount of attention to the articulation and codification of knowledge, but held that an essential part of knowledge was not articulable and that such tacit knowing fundamentally guided all action. Despite the epistemological impossibility of fully "converting" knowledge between its tacit and explicit states, and the multitude of forms that knowledge may take (Winter 1987; Blackler 1995; Zander and Kogut 1995; Gourlay 2004), people nevertheless give partial but consequential expression to their knowing all the time. Shifting attention to the underlying processes by which certain aspects of knowing are continually, meaningfully, and consequentially communicated allows us to avoid intractable epistemological debates regarding the nature and boundaries of knowledge states, while keeping the discrimination these debates exhibit in mind.

6. Highly codified knowledge, though nominally explicit, may ironically come to be ratified, taken for granted, and therefore in some sense tacit.

7. Dougherty (1992a,b; Dougherty and Hardy 1996), for example, has explored the knowledge processes associated with cross-boundary, coordinative work that distinguishes successful product development efforts.

8. *Oxford English Dictionary.*

9. There is the connection here to the noticing, bracketing, and labeling described in sensemaking (Weick et al. 2005; Stigliani and Ravasi 2012) that serves as a precursor to knowledge articulation.

10. *Macmillan Dictionary*, online.

11. This definition was developed collaboratively with Paul Carlile.

12. My field data show that strategic actors routinely failed to successfully communicate issues, concerns, requests, etc. to their interlocutors.

13. Note the correspondence here with Latour's (1987) depiction of enrollment.

14. Existing technologies often evolve to taken-for-granted status where support is no longer problematic.

15. Tools are necessary for stamping out the metal parts used to assemble automobiles. Tom's comment alludes to the fact that there are two more stages of tooling: refinement of preliminary tools (known as "soft tools") before the final "hard tools" are produced in order for the car to go into full assembly-line production.

16. The "stipple" refers to the relative bumpiness or smoothness of a surface.

17. Articulation devices included verbal, written, or physical representations such as analogies, metaphors, stories, physical objects, PowerPoint, and informal sketches that facilitated knowledge articulation.

18. As an engineer and the manager responsible for the steel stamping of the car, Brad already was predisposed to place great importance on the integrity of the G5's visual representations. The importance that the Virtual Build database held for Brad was in part rooted in the importance of visual representation to automotive design and engineering work more broadly (Henderson 1998). At NewCar, it was quite common for design meetings to wait silently for several minutes while the 3-D representations of the automotive parts under consideration were brought up on a projected computer display. In the period of my observation, virtual, geometrically accurate, two-dimensional designs had already supplanted the physical drawing work that for more than a half century had been executed on physical drafting boards and vellum paper, a design process that the older designers still remembered vividly. Now 3-D technology was supplanting 2-D.

19. See Chapter 7 for an in-depth consideration of these issues.

20. Goffman allowed for more than two teams but emphasized the two-team format.

21. Curation inevitably leads back to considerations of timing, or saying the right thing at the right time, given the context and interlocutors present. Social skill speaks to this issue.

22. This insight was provided by Michael Cohen.

23. Boisot and Child define "abstraction" as a "reduction in the number of categories to which data needs to be assigned for a phenomenon to be apprehended" (1999, 239).

24. There is a tension between the move between complex and simple where a move to simplify can simultaneously and paradoxically lead to a more complex view of a situation or problem. An illustration of that paradox is what Weick refers to as the creation of a cause map, a picture of how someone perceives elements to be causally or sequentially related—the kind of sketch made at corporate whiteboards daily. Weick indicates, "A cause map itself is [a] simplification, even though it does help to complicate how an individual examines his organization" (1979, 261).

25. In his essay "Art as Technique," Shklovsky describes how literature allows people to see things with a fresh perspective, breaking away from taken-for-granted, automatic understandings: "Habitualization devours works, clothes, furniture, one's wife, and the fear of war. And art exists that one may recover the sensation of life; it exists to make one feel things, to make the stone stony. The purpose of art is to impart the sensation of things as they are perceived and not as they are known" (1988, 18–19).

26. Smith writes, "The ideal of explanation is occasioned by surprise and seeks to reduce surprise. That is to say, explanation has as its goal the reduction of that which at first appears unknown to an instance of that which is already known" (1985, 57–58).

27. The oscillation between and coexistence of defamiliarizing and familiarizing suggest a liminal space between tacit and explicit knowing which I refer to as a zone of emergent knowledge. Familiarizing efforts are important to the extent that they make unfamiliar knowledge more available. These provisional, middle-range familiarizing efforts can be contrasted with more complete naturalization of an object, idea, or person. According to Bowker and Star, "Naturalization means stripping away the contingencies of an object's creation and its situated nature. A naturalized object has lost its anthropological strangeness. It is in that narrow sense desituated—members have forgotten the local nature of the object's meaning or the actions that go into maintaining and recreating its meaning" (2000, 299).

28. Knowledge articulation displays a range of defamiliarizing intensity. In some cases, defamiliarization involves a salient break from taken-for-granted understanding and even jarring confrontations. In other cases, defamiliarization works more subtly to raise awareness or, as Shklovsky (1988) described it, "to make the stone more stony."

29. The presence of the prototype armrest as an accompanying articulation device goes a long way toward making these exchanges syntactical.

30. Articulation artifacts for this study were routinely recovered from trash cans and tabletops.

Chapter 3

A portion of the theory presented here is drawn from Obstfeld (2012).

1. While the creative project construct's denotation of emergent action, consistent with the pragmatist tradition, could apply to individual action, my organizational focus emphasizes interdependent action among multiple actors.

2. Dewey's work emphasizes individual conduct. Though closely tied to Dewey's tripartite model, my focus, given my emphasis on organizational innovation, is on the collective applications of creative projects and organizational routines.

3. I leave out for simplicity the observation that airplane construction is a function of many constituent routines, and that an experienced airplane manufacturer might eventually learn how to routinize the process by which it migrates from one assembly line production setup to another.

4. There are many alternative routes by which organizations might confront such a transition. See, for example, Adler et al.'s (1999) depiction of Toyota's "refection-review," or *hansei* process, a routine created to improve the automotive model change process, that is, a routine to facilitate the transition from one manufacturing routine to another.

5. The word "project" is a translation of the German *entwurf*, as used by Schütz, Heidegger, and Kant.

6. It is also important to situate this treatment of projects within the well-developed literature on project management, which makes a distinction between ongoing, repetitive operations and projects. The project is defined by one textbook as "a temporary endeavor undertaken to create a unique product or service" (Duncan 1996, 4). Although they might be temporary and unique in status, some projects—as suggested above—in practice display the repetitiveness of routines. A given product devel-

opment effort, despite being widely labeled as a "project," might be pursued very similarly to the product development efforts that preceded it, and would therefore be categorized as more akin to a routine. The adjective "creative" is used herein to denote an alternative, less repetitive project form.

7. One German word that Schumpeter used in his second draft (1926) of *Theory of Economic Development* (1934) was *Bahn*, which might be translated most simply as "path" or movement in a particular direction.

8. In an earlier treatment of the creative project (Obstfeld 2012), I suggested that the creative project and organizational routine had similar ostensive and performative dimensions, dimensions introduced by Feldman and Pentland (2003) to describe different aspects of organizational routines. For Pentland and Feldman the ostensive involves "the abstract or generalized pattern of the routine" that "participants use . . . to guide, account for, and refer to specific performances of the routine" (2005, 796). There is a loose initial correspondence between this ostensive aspect of the organizational routine and the creative project's trajectory projection, which, as I alluded to above, motivates and guides action. Similarly, there is a rough analogue between what Feldman and Pentland (2003) refer to as the performative aspect of routines, the "actual performances by specific people, at specific times, in specific places," and the actual performance of the creative project itself. A closer examination, however, suggests limitations to locating an analogue for the ostensive and performative aspects of the organizational routine in the creative project. Both the project and organizational routine share a trajectory projection, an abstract conceptualization that guides action that may vary among the various participants within or surrounding a routine; the ostensive aspect of the routine, however, according to Feldman and Pentland (2003), involves a typification, that is, a generalized abstraction of the routine's characteristics that reflects participants' or observers' experience of its repetitive functioning over time. Strictly speaking, this typification would correspond not with the routine's trajectory projection but only with its outcome, i.e., the routine's endpoint. Even if some organizational actors use the routine's outcome to reference the routine, the ostensive, as used by Feldman and Pentland, would emphasize the ostensive aspect's correspondence with a more complex representation of the entire routine's functioning. In the case of the creative project, on the other hand, because a trajectory involves newly emerging action, the trajectory projection concerns an outcome that is yet to be accomplished, and a typification based on past executions of the trajectory therefore does not exist. It should be noted, however, that as action within a creative project accumulates, an emergent history and narrative comes to be associated with a given trajectory projection, and a representation comparable to a routine's ostensive aspect could be said to exist.

9. Note how this framing is reminiscent of Joas's description of the pragmatist orientation provided at the start of this chapter: "Hypotheses are put forward: suppositions about new ways of creating bridges between the impulses to action and the given circumstances of the situation. Not all such bridges are viable" (1996, 133). In the same spirit, one might add that not all such bridges are visible or apparent at any point on the path.

10. When I invoke "foolishness," I refer to March's (1971) paper "The technology of foolishness," in which he proposes an alternative to rational action that pursues purposes, decisions, and action aligned behind well-established objectives. Foolishness in the form of playfulness or improvisation leads to action that "might help . . . in a small way to develop those unusual combinations of attitudes and behaviors" (265), which lead to new ideas, many of which may not be good but may align with the spirit of exploration (as opposed to exploitation).

11. Actor network theory (Callon 1986; Latour 2005) presents a compatible perspective.

12. The "garbage can" model of organizing (Cohen et al. 1972) views certain forms of organizing as involving different combinations of problems, solutions, participants, and choice opportunities. In this view, "organizational anarchies" are characterized by "choices looking for problems, issues and feelings, looking for decision situations in which they might be aired, solutions looking for issues to which they might answer, and decision makers looking for work" (1). In the view presented here, I would expand the model's view of choice (or decision) to go beyond formal "occasions of choice" (3) to include more frequent informal choices about what to do next in a given action trajectory

13. One assumption here is that the routines and projects under consideration are of roughly equal magnitude in terms of participation. This is meant to rule out comparisons of routines and projects of significantly different magnitude or scope.

14. Note that I infer more than I know about this episode.

Chapter 4

1. For methods, see Appendix 1. See also, as a supplement to the ethnographic data, corroborating survey findings from NewCar in Appendix 2.

2. For those too young to have been exposed to this technology, "manual shifting" involves a clutch pedal on the floor, next to the brake, and a stick for selecting gears manually—as opposed to automatically—most often mounted on the console between the front seats.

3. This status arrangement is not inherent to automotive design. In Japan, the engineering work associated with manufacturing (i.e., production engineering) has a higher status than design engineering (Sobek 1997).

4. EDRM is a pseudonym.

5. As noted in earlier chapters, my fieldwork took place at NewCar (a pseudonym), a division of a major automotive manufacturer I refer to as AllCar. NewCar had over one thousand employees, approximately four hundred of whom were dedicated to the design of a new vehicle that I refer to as the G5.

6. Figure 4.3 shows the placement of the four outermost shifter positions (i.e., first, second, fifth, and reverse gear) for "move 24," with the interference circled.

7. In contrast to this chapter's emphasis on a general pattern for emergent innovative action over time consisting of the interaction of knowledge and network processes, slippage denotes the absence of such processes.

8. As noted, the team met three times a week, so in that month I attended twelve meetings.

9. These errors were sometimes technical in nature but were ones that I, as a nonengineer, was able to detect after a week of close observation and recording of field notes.

10. Henry's sign itself suggests a more subtle form of defamiliarization. The sign's formal, public description of informal engineering resistance to NVH makes that resistance "strange" and potentially subject to change. A part engineer who visited his cubicle could not help but become more aware of, and potentially rethink, his or her habituated forms of resistance to Henry's NVH initiatives.

11. Henry also prepared a six-page memo with detailed calculations that made the case that another part would not have satisfactory NVH characteristics. During crunch team observations, I did not observe any other participant prepare such reports.

12. As noted in Chapter 2, I used this alternative formulation from Mead (1934) to avoid conceptual baggage associated with role theory.

13. An enlarged rubber isolator on the manual shifter was the other NVH design change for which Henry was advocating within the crunch team.

14. Survey data confirmed the larger size and greater number of structural holes in Henry's social network.

15. NewCar engineers generally had college or master's degrees, while designers often had little or no college education.

16. I was surprised to learn from the designers' manager, nearly a month after their arrival, that he had deployed Alex and Joe to the crunch team after recognizing the team's struggles.

17. I was not specifically tracking Alex and Joe until this meeting, but logged the names of every meeting's participants.

18. Joe was asking, somewhat incredulously, whether there was still a bidding process at this late stage in the design process.

19. Note *alter articulation* is a variant of *perspective articulation* discussed in Chapter 6.

20. I too did not know who Alex and Joe were at the point that they initiated their various interruptions. Alex and Joe's interjections disrupted not only the team's design routines but also my relatively straightforward note-taking process. My choice to follow them closely after the meeting—to them, an odd outcome—was motivated by a desire to figure out who they were. That is how I came to record the dialogue reported here.

21. Whereas Sam had a barely adequate grasp of the 3-D software, Alex and Joe had extensive experience with it.

22. Joe was the manager who admired Rick's design work in Chapter 2.

23. To the extent that frames allow individuals "to locate, perceive, identify, and label" (Goffman 1974, 21), the master cross-sections constituted a frame.

24. Latour (1986) remarked on the power of rendering more complex, multidimensional data into two-dimensional inscriptions.

25. A "typical section" was another name for a two-dimensional master cross-section.

26. Note that Joe was not an engineer but a designer employed in the Engineering division. Despite the difference in status within engineering, the stylist didn't make a distinction between designers and engineers.

Chapter 5

1. Field observations of the PPP routine were conducted at the same time as those provided in Chapter 5. I studied the PPP routine through extensive interviews with key participants, direct observation of prototype builds and prototype part purchasing in progress, and observation of meetings in which the implications of prototype test results and subsequent build strategies were discussed. This data collection was supplemented by direct observation of and interviews with engineers who were designing, revising, and ordering automotive parts as part of the ongoing G5 prototype builds.

An introduction by the G5 program manager, Dan, provided the opportunity to meet the cross-divisional AllCar creative project group and directly observe much of its effort to change the existing routine. Over a five-month period, I attended twelve of the group's meetings, more than a dozen less formal gatherings, and a corporate-wide three-day retreat convened to evaluate changing the PPP routine, which occurred in the third month of the group's efforts. In the second month, the group agreed to allow me to tape all of its meetings (the audio tapes were subsequently transcribed), and I was placed on its e-mail distribution list. I maintained continuous, informal contact with several of the group members throughout the observation period.

At about the same time that I was observing the AllCar creative project, Dan determined that he would pursue the more immediate creation of a G5 PPP unit. I refer to this effort as the "G5 creative project." Over a five-month period, I conducted ten formal interviews with Dan, all of which were taped and transcribed, as well as interviews with several other people associated with the initiative. I also observed a series of informal meetings through which Dan successfully created a new PPP team devoted to managing the G5's prototype builds. For both the PPP routine and the two creative project trajectories, I wrote narratives that pulled together the information gained from field observations, meetings, interviews, and project-related documents. My approach to data analysis is described in Appendix 1.

2. MROB is a pseudonym for the acronym by which the building was known.

3. While the five dimensions of knowledge articulation—back stage/front stage; complex/simple; past, present, and future; familiar/unfamiliar; and laying down markers—as well as knowledge articulation's connection to transfer, translation, and transformation were also in evidence, I focus on fixing, pitching, and scheming in this case.

Chapter 6

1. Power (Pfeffer 2010), intelligence, and network structure account for others, but here I emphasize how entrepreneurs get new things done by leveraging one or more of three possible spheres.

2. Knowledge articulation would also more broadly include framing within organizations (Kaplan 2008) and social movements (Snow et al. 1986).

3. It is technically possible to identify "ties" that involve no reciprocal exchange, but the dyadic interact suggests a minimal relational requirement.

4. Blumer and Morrione similarly observe that "the interaction between two individuals . . . is the prototype of human social interaction in general" (2004, 23).

5. Note that Mead (1934) and Weick (1979) are referring to the same three-step process with different terminology: Mead's use of "triadic" (which he also referred to as the "triadic matrix") concerns the three-step process. Weick's double interact features the same three-step process but is labeled "double," where the first interact is the response to the initiating gesture and the second interact is the initiating actor's response to that response.

6. Note Lave's (1991) construct of legitimate peripheral participation, which addresses how membership and expertise are fluid and evolve over time.

7. Riffing may be seen as a subcategory of role-playing. Role-playing has a broader meaning that may refer to structured exercises in connection with training, dispute resolution, therapy, acting, or decision-making-alternative generation.

8. The effectiveness that ventriloquation displayed suggests that the idea of a tool kit (Swidler 1986; Wertsch 1991) be expanded to include the different social languages at the entrepreneur's disposal.

9. Cooren (2010) points out that such multivocality may not only lead to the enhanced capacity to get new things done but also have an ethical dimension by registering, giving voice to, and reconciling multiple perspectives.

10. In July 2015, Dan read this chapter and my account of his activity. He indicated that the chapter had accurately captured his process, with one caveat. He indicated that his social process was "contemporaneous not extemporaneous," in the sense that he never spent any time off-line "preplanning" before engaging socially. He indicated, "I do recognize that there is intentionality. I do recognize what I am doing. I do it contemporaneously, in real time; I'm not scripting it [in advance]."

Chapter 7

1. Future interesting applications of the BKAP model might include: one means for resolving the microfoundations debate (Barney and Felin 2013; Felin and Foss 2005; Winter 2011); the nonroutine origins of dynamic capability (Teece and Pisano 1994; Eisenhardt and Martin 2000; Teece 2012, 2014; Winter 2003); a fast-growing literature tracing how brokerage orientations at the firm level help account for dynamic new forms of supply chain management (Choi and Wu 2009; Li and Choi 2009); new frontiers of sensemaking (Weick et al. 2005; Stigliani and Ravasi 2012; Strike and Rerup 2016); ambidexterity as involving a tension between routine and nonroutine action (Gupta, Smith, and Shalley 2006; Raisch and Birkinshaw 2008; Smith and Tushman 2005); how transactive memory systems (Lee et al. 2014) emerge from nonroutine situations; the relationship between social skill and emotional intelligence (Salovey and Mayer 1990; Boyatzis, Stubbs, and Taylor 2002); a process explanation of job crafting (Berg, Wrzesniewski, and Dutton 2010); and social skill and brokerage action orientations of "givers" versus "takers" (Grant 2013).

2. In part due to the turmoil associated with the Russian revolution, the Ballets Russes never actually performed in Russia or had any official ties to that country.

3. Per draft of chapter "Entrepreneurial Groups" in Ruef (2009).

4. After Diaghilev's death, the Ballets Russes was dissolved and reformed as the "Original Ballet Russe" in 1932 and the Ballet Russe de Monte Carlo in 1938.

5. The Ballets Russes' broad impact is even more remarkable when viewed alongside its frequent preoccupation with simply surviving; it dissolved immediately after Diaghilev's death with significant debt. The broad impact emerging from this resource-starved entity is strikingly similar to the industry-wide innovation that emerged from NewCar, which was similarly consumed with simply surviving and competing with much larger automotive manufacturers.

6. It is this looser form of "control" exerted by the impresario that Stovel and Shaw (2012) reference because of its catalytic effect, in the sense that the broker alters the rate of interaction among other actors. The full extension of this catalytic potential is referenced by Padgett and Powell's (2012) use of autocatalysis, which they argue goes beyond traditional conceptualizations of diffusion. Padgett and Powell suggest that "autocatalytic networks are networks of transformation, not networks of mere transmission," and similarly, that "diffusion should be reconceptualized from mimicry to chain reactions" (2012, 9). The transition from Stovel and Shaw's more local account of the *tertius iungens* broker's impact (i.e., catalysis), to Padgett and Powell's account of social processes that spread well beyond their local origins (i.e., autocatalysis), however, contrasts a phenomenon where actors remain "on stage" to one in which they are nudged to the periphery, in favor of "structured social interaction, endogenous or not, chains of rules and protocols [that] assemble themselves via autocatalytic growth into technologies . . . , markets . . . , and language communities" (Padgett and Powell 2012, 9).

My account of brokerage emphasizes brokers as agents on stage but acknowledges that social processes may amplify to a point where individual intent no longer has substantive impact—Padgett and Powell's central point. My account also stresses more overt agency than the sphinxlike multivocality and "noncommittal actions that keep future lines of action open" (Padgett and Powell 2012, 24) found in "robust action" (Padgett and Ansell 1993). Robust action and the more active orchestration suggested by the BKAP model, however, are not mutually exclusive but can be seen as options in the broker's toolkit that he or she employs in different mixes on the basis of strategic choices about the level of overt engagement that is appropriate to a given situation.

7. Schumpeter's five fundamental forms of new combinations include "a new good," "a new method of production," "a new market," "a new source of supply of raw materials," and "the carrying out of a new organization of any industry" (1934, 66).

8. Gamson et al.'s (1982) account was drawn from Kraus (1947) and Fine (1969). Kraus edited the *Flint Auto Worker* for the fledgling United Automobile Workers during the events described.

9. The word "began" is in quotes to recognize that there are multiple alternative beginnings that one could point to in demarcating any collective action trajectory.

10. Kraus (1947) does provide extensive details of this trajectory in his book.

11. Clemens (2005) distinguishes between the manifestation of such coordination without authority within firms and in organizational fields.

12. Framing (e.g., Snow et al. 1986) addresses the intentional, interpretive work that aligns interests and generates participation (McAdam et al. 1996; Campbell 2005; Snow 2006). Advocates use frames to organize individual and group experiences in ways that help new movements emerge. Framing processes are emergent, interpretive schemes that allow individuals "to locate, perceive, identify, and label" (Goffman 1974, 21) occurrences in their daily experience. My emphasis on knowledge articulation suggests a more granular instantiation of communicative processes that framing addresses. Knowledge articulation concerns communicative activity within a broad set of purposeful, microsocial episodes or encounters (Goffman 1961; Gamson et al. 1982) from which effective frames might arise (Kaplan 2008). A small subset of articulated knowledge emerges as frames that are instrumental in organizing or influencing the collective experience in some enduring way. Frames are the salient and influential meanings that are retained from multiple variation, selection, and retention cycles of knowledge articulation (Campbell 1960; Weick 1969; Weick et al. 2005).

13. Hedström and Swedberg (1988, 5) specifically reference the *tertius gaudens* as an example of a mechanism. The absence of *tertius gaudens* in McAdam et al. (2001) and Tilly and Tarrow's (2007) approach is addressed further below.

14. In this respect, I return to the observation made in Chapter 3 of a creative project built or stacked on routines and Strauss's (1993) observation that even revolutionary actions involve a "repertoire of routines."

15. Ghonim (2012) points out, in what could be understood in political opportunity structure terms, that the number of Egyptian internet users increased from 1.5 million to 13.6 million between 2004 and 2008.

16. Ghonim indicates, "As soon as someone 'likes' a page, Facebook considers the person and the page to be 'friends.' So if the 'admin' of the page writes a post on the 'wall,' it appears on the walls of the page's fans. This is how ideas can spread like viruses. A particular post can appear on the users' walls to be viewed thousands, or even millions, of times. In the case of groups, however, users have to access the group to remain updated; no information is pushed out to them" (2012, 43).

17. The language of rapid, repeated cycles of experimentation and learning could be lifted from a Lean Startup entrepreneurship manual (e.g., Ries 2011; Blank 2013), with the only distinction being the novelty, at least at the outset, of Ghonim's collective action techniques. The techniques now associated with Lean Startup feature a fairly standardized action repertoire that includes split testing and minimum viable prototypes.

18. In exploring organizing along the analogue-digital divide, I encountered a fascinating debate between Malcolm Gladwell (2010) and Clay Shirky (2008) as to the impact of social media on modern social revolutions. Shirky argues, "Digital networks have acted as a massive positive supply shock to the cost and spread of information, to the ease and range of public speech by citizens, and to the speed and scale of group coordination" (2011, 154), a position that Gladwell questions: "What evidence is there that social revolutions in the pre-Internet era suffered from a lack of cutting-edge communications and organizational tools? [For Shirkey's] argument to be anything close to persuasive, he has to convince readers that in the absence of social media, those uprisings would not have been possible" (2010, 153). Citing McAdam's (1990) research on Freedom Summer, Gladwell emphasizes strong "analogue" ties, not social media, as a key factor that sustains activism. McAdam's point was a little more nuanced: "It is a strong subjective identification with a particular identity, reinforced by organizational or individual ties, that is especially likely to encourage participation" (McAdam and Paulsen 1993, 659). The lines of the Gladwell-Shirky debate return us to a central theme of the book: in getting new things done, we need to attend not only to structure—either analogue or digital, whether as a digital pipe for information (i.e., Shirky) or in terms of strong ties that better secure movement participation (i.e., Gladwell)—but to the closely related but distinct *process* whereby people and interests coalesce and expand.

19. The opportunity and challenges associated with this population can't be overstated. Research shows that students who begin in two-year institutions/community colleges are far less likely to attain a bachelor's degree than those that begin in a four-year institution (Kurlaender and Flores 2005). In addition, even students who do make the transfer from a two-year to four-year institution are less likely to complete a bachelor's degree, and students from lower socioeconomic-status backgrounds are more likely to pursue this pathway (e.g., Goldrick-Rab 2006).

References

Adler, P. S., B. Goldoftas, and D. I. Levine. 1999. "Flexibility versus efficiency? A case study of model changeovers in the Toyota production system." *Organization Science* 10(1): 43–68.

Adler, P. S., and D. Obstfeld. 2007. "The role of affect in creative projects and exploratory search." *Industrial and Corporate Change* 16(1): 19–50.

Ahuja, G. 2000. "Collaboration networks, structural holes, and innovation: A longitudinal study." *Administrative Science Quarterly* 45(3): 425–455.

Alvarez, S. A., and J. B. Barney. 2005. "How do entrepreneurs organize firms under conditions of uncertainty?" *Journal of Management* 31(5): 776–793.

———. 2007. "Discovery and creation: Alternative theories of entrepreneurial action." *Strategic Entrepreneurship Journal* 1(1–2): 11–26.

Alvarez, S. A., J. B. Barney, and P. Anderson. 2013. "Forming and exploiting opportunities: The implications of discovery and creation processes for entrepreneurial and organizational research." *Organization Science* 24(1): 301–317.

Alvarez, S. A., and L. W. Busenitz. 2001. "The entrepreneurship of resource-based theory." *Journal of Management* 27(6): 755–775.

Alvesson, M., and D. Kärreman. 2001. "Odd couple: Making sense of the curious concept of knowledge management." *Journal of Management Studies* 38(7): 995–1018.

Amabile, T. M. 1996. *Creativity in Context: Update to "The Social Psychology of Creativity."* Boulder, Colo.: Westview Press.

Aral, S., and M. Van Alstyne. 2011. "The diversity-bandwidth trade-off." *American Journal of Sociology* 117(1): 90–171.

Argote, L. 2012. *Organizational Learning: Creating, Retaining and Transferring Knowledge.* New York: Springer Science & Business Media.

Argote, L., and H. R. Greve. 2007. "A behavioral theory of the firm—40 years and counting: Introduction and impact." *Organization Science* 18(3): 337–349.

Argyris, C., and D. A. Schön. 1978. *Learning Organizations: A Theory of Action Perspective.* Reading, Mass.: Addison-Wesley.

Axelrod, R. 1984. *The Evolution of Cooperation.* New York: Basic Books.

Baker, T., A. S. Miner, and D. T. Eesley. 2003. "Improvising firms: Bricolage, account giving and improvisational competencies in the founding process." *Research Policy* 32(2): 255–276.

Baker, T., and R. E. Nelson. 2005. "Creating something from nothing: Resource construction through entrepreneurial bricolage." *Administrative Science Quarterly* 50(3): 329–366.

Baker, W. E., and R. R. Faulkner. 1991. "Role as resource in the Hollywood film industry." *American Journal of Sociology* 97(2): 279–309.

Bakhtin, M. M. 1981. *The Dialogic Imagination: Four Essays by MM Bakhtin.* Edited by M. Holquist and C. Emerson. Translated by M. Holquist. Austin: University of Texas Press.

Bargh, J. A. 1989. "Conditional automaticity: Varieties of automatic influence in social perception and cognition." *Unintended Thought* 3: 51–69.

Barley, S. R. 1996. "Technicians in the workplace: Ethnographic evidence for bringing work into organizational studies." *Administrative Science Quarterly* 41(3): 404–441.

Barney, J., and T. Felin. 2013. "What are microfoundations?" *Academy of Management Perspectives* 27(2): 138–155.

Beaumont, C. W. 1940. *The Diaghilev Ballet in London: A Personal Record*. London: A & C Black.

Bechky, B. A. 2003. "Sharing meaning across occupational communities: The transformation of understanding on a production floor." *Organization Science* 14(3): 312–330.

———. 2006. "Gaffers, gofers, and grips: Role-based coordination in temporary organizations." *Organization Science* 17(1): 3–21.

Becker, M. C. 2004. "Organizational routines: A review of the literature." *Industrial and Corporate Change* 13(4): 643–678.

Becker, M. C., T. Knudsen, and J. G. March. 2006. "Schumpeter, Winter, and the sources of novelty." *Industrial and Corporate Change* 15: 353–371.

Beckman, C. M., and P. R. Haunschild. 2002. "Network learning: The effects of partners' heterogeneity of experience on corporate acquisitions." *Administrative Science Quarterly* 47(1): 92–124.

Benner, P. 1994. "The role of articulation in understanding practice and experience as sources of knowledge in clinical nursing." In *Philosophy in an Age of Pluralism: The Philosophy of Charles Taylor in Question*, 136–155. Cambridge: Cambridge University Press.

Berg, J. M., A. Wrzesniewski, and J. E. Dutton. 2010. "Perceiving and responding to challenges in job crafting at different ranks: When proactivity requires adaptivity." *Journal of Organizational Behavior* 31(2–3): 158–186.

Berliner, P. 1994. *Thinking in Jazz: The Ultimate Art of Improvisation*. Chicago: University of Chicago Press.

Birnholtz, J. P., M. D. Cohen, and S. V. Hoch. 2007. "Organizational character: On the regeneration of Camp Poplar Grove." *Organization Science* 18(2): 315–332.

Bizzi, L., and A. Langley. 2012. "Studying processes in and around networks." *Industrial Marketing Management* 41(2): 224–234.

Blackler, F. 1995. "Knowledge, knowledge work and organizations: An overview and interpretation." *Organization Studies* 16(6): 1021–1046.

Blank, S. 2013. *The Four Steps to the Epiphany*. Pescadero, Calif.: K&S Ranch Press.

Blumer, H., and T. J. Morrione. 2004. *George Herbert Mead and Human Conduct*. Walnut Creek, Calif.: AltaMira Press.

Boisot, M. 1995. *Information Space*. New York and London: Routledge.

Boisot, M., and J. Child. 1999. "Organizations as adaptive systems in complex environments: The case of China." *Organization Science* 10(3): 237–252.

Boltanski, L., and E. Chiapello. 2005. "The new spirit of capitalism." *International Journal of Politics, Culture, and Society* 18(3–4): 161–188.

———. 2007. *The New Spirit of Capitalism*. New York: Verso.

Bonacich, E. 1973. "A theory of middleman minorities." *American Sociological Review* 38(5): 583–594.

Bourdieu, P. 1996. *The State Nobility: Elite Schools in the Field of Power*. Oxford: Polity.

Bowker, G. C., and S. L. Star. 2000. *Sorting Things Out: Classification and Its Consequences*. Cambridge, Mass.: MIT Press.

Boyatzis, R. E., E. C. Stubbs, and S. N. Taylor. 2002. "Learning cognitive and emotional intelligence competencies through graduate management education." *Academy of Management Learning and Education* 1(2): 150–162.

Brown, J. S., and P. Duguid. 2001. "Knowledge and organization: A social-practice perspective." *Organization Science* 12(2): 198–213.

Brown, S. L., and K. M. Eisenhardt. 1997. "The art of continuous change: Linking complexity theory and time-paced evolution in relentlessly shifting organizations." *Administrative Science Quarterly* 42(1): 1–34.

Burgelman, R. A. 1991. "Intraorganizational ecology of strategy making and organizational adaptation: Theory and field research." *Organization Science* 2(3): 239–262.

———. 1994. "Fading memories: A process theory of strategic business exit in dynamic environments." *Administrative Science Quarterly* 39(1): 24–56.

Burns, T. E., and G. M. Stalker. 1961. *The Management of Innovation*. Academy for Entrepreneurial Leadership Historical Research Reference in Entrepreneurship, University of Illinois, Urbana-Champaign.

Burt, R. S. 1978. "Cohesion versus structural equivalence as a basis for network subgroups." *Sociological Methods and Research* 7(2): 189–212.

———. 1992. *Structural Holes*. Cambridge, Mass.: Harvard Business School Press.

———. 1997. "The contingent value of social capital." *Administrative Science Quarterly* 42(2): 339–365.

———. 2000. "The network structure of social capital." *Research in Organizational Behavior* 22: 345–423.

———. 2002. "The social capital of structural holes." In *The New Economic Sociology: Developments in an Emerging Field*, 148–190. New York: Russell Sage Foundation.

———. 2004. "Structural holes and good ideas." *American Journal of Sociology* 110(2): 349–399.

———. 2005. *Brokerage and Closure*. New York: Oxford University Press.

———. 2010. *Neighbor Networks: Competitive Advantage Local and Personal*. Oxford: Oxford University Press.

———. 2013. "Network structure of advantage." Working paper, University of Chicago, Booth School of Business.

Burt, R. S., M. Kilduff, and S. Tasselli. 2013. "Social network analysis: Foundations and frontiers on advantage." *Annual Review of Psychology* 64: 527–547.

Burt, R., and J. Merluzzi. 2016. "Network oscillation." *Academy of Management Discoveries* 2: 368–391.

Burt, R. S., J. L. Merluzzi, and J. G. Burrows. 2013. "Path dependent network advantage." In *Proceedings of the 2013 Conference on Computer Supported Cooperative Work*, 1–2. New York: ACM Digital Library.

Callon, M. 1986. "Some elements of a sociology of translation: Domestication of the scallops and the fishermen of St. Brieuc Bay." *Power, Action, and Belief: A New Sociology of Knowledge* 32: 196–223.

Campbell, A. 1960. "Surge and decline: A study of electoral change." *Public Opinion Quarterly* 24(3): 397–418.

Campbell, J. L. 2005. "Where do we stand?" In *Social Movements and Organization Theory*, 41–48. Edited by G. F. Davis, D. McAdam, W. R. Scott, and M. N. Zald. Cambridge: Cambridge University Press.

Cardinal, L. B., S. B. Sitkin, and C. P. Long. 2004. "Balancing and rebalancing in the creation and evolution of organizational control." *Organization Science* 15(4): 411–431.

Carlile, P. R. 2002. "A pragmatic view of knowledge and boundaries: Boundary objects in new product development." *Organization Science* 13(4): 442–455.

———. 2004. "Transferring, translating, and transforming: An integrative framework for managing knowledge across boundaries." *Organization Science* 15(5): 555–568.

Carlile, P. R., and E. S. Rebentisch. 2003. "Into the black box: The knowledge transformation cycle." *Management Science* 49(9): 1180–1195.

Casson, M. 1982. *The Entrepreneur: An Economic Theory*. Totowa, N.J.: Rowman & Littlefield.

———. 2010. "Entrepreneurship, business culture and the theory of the firm." In *Handbook of Entrepreneurship Research*, 249–271. New York: Springer.

Cattani, G., and S. Ferriani. 2008. "A core/periphery perspective on individual creative performance: Social networks and cinematic achievements in the Hollywood film industry." *Organization Science* 19(6): 824–844.

Choi, T. Y., and Z. Wu. 2009. "Taking the leap from dyads to triads: Buyer-supplier relationships in supply networks." *Journal of Purchasing and Supply Management* 15(4): 263–266.

Clemens, E. S. 2005. "Two kinds of stuff." In *Social Movements and Organization Theory*, 351. Edited by G. F. Davis, D. McAdam, W. R. Scott, and M. N. Zald. Cambridge: Cambridge University Press.

Cohen, M. 2007. "Reading Dewey: Reflections on the study of routine." *Organization Studies* 28(5): 773–786.

Cohen, M. D., and P. Bacdayan. 1994. "Organizational routines are stored as procedural memory: Evidence from a laboratory study." *Organization Science* 5(4): 554–568.

Cohen, M. D., R. Burkhart, G. Dosi, M. Egidi, L. Marengo, M. Warglien, and S. Winter. 1996. "Routines and other recurring action patterns of organizations: Contemporary research issues." *Industrial and Corporate Change* 5(3): 653–698.

Cohen, M. D., J. G. March, and J. P. Olsen. 1972. "A garbage can model of organizational choice." *Administrative Science Quarterly* 17(1): 1–25.

Cohen, W. M., and D. A. Levinthal. 1990. "Absorptive capacity: A new perspective on learning and innovation." *Administrative Science Quarterly* 35(1): 128–152.

Coleman, J. S. 1988. "Social capital in the creation of human capital." *American Journal of Sociology* 94: S95–S120.

———. 1990. *Foundations of Social Theory*. Cambridge, Mass.: Belknap.

Collins, H. M. 1985. *Changing Order: Replication and Induction in Scientific Discourse.* Chicago: University of Chicago Press.

Collins-Dogrul, J. 2012. "*Tertius iungens* brokerage and transnational intersectoral cooperation." *Organization Studies* 33(8): 989–1014.

Cook, G. A. 1993. *George Herbert Mead: The Making of a Social Pragmatist.* Champaign: University of Illinois Press.

Cook, S. D., and J. S. Brown. 1999. "Bridging epistemologies: The generative dance between organizational knowledge and organizational knowing." *Organization Science* 10(4): 381–400.

Cooper, H. 2009. "US officials get a taste of Pakistanis' anger at America." *New York Times*, August 19.

Cooren, F. 2010. *Action and Agency in Dialogue: Passion, Incarnation, and Ventriloquism.* Vol. 6. Philadelphia: John Benjamins.

Cooren, F., and S. Sandler. 2014. "Polyphony, ventriloquism, and constitution: In dialogue with Bakhtin." *Communication Theory* 24(3): 225–244.

Cyert, R. M., and J. G. March. 1963. *A Behavioral Theory of the Firm.* Englewood Cliffs, N.J.: Prentice-Hall.

Czarniawska, B., and B. Joerges. 1996. "Travels of ideas." In *Translating Organizational Change.* Edited by B. Czarniawska and G. Sevon. Berlin: Walter de Gruyter.

Davis, G. F., D. McAdam, W. R. Scott, and M. N. Zald, eds. 2005. *Social Movements and Organization Theory.* Cambridge: Cambridge University Press.

Davis, G. F., and M. N. Zald. 2005. "Social change, social theory, and the convergence of movements and organizations." In *Social Movements and Organization Theory*, 335–350. Edited by G. F. Davis, D. McAdam, W. R. Scott, and M. N. Zald. Cambridge: Cambridge University Press.

Davis, J. P. 2010. "Agency and knowledge problems in network dynamics: Brokers and bridges in innovative interorganizational relationships." Working paper, MIT Sloan School of Management.

———. 2011. "Network agency problems: Reconceptualizing brokerage as a barrier to embedded relationships." Working paper, MIT Sloan School of Management.

Davis, J. P., and K. M. Eisenhard. 2011. "Rotating leadership and collaborative innovation recombination processes in symbiotic relationships." *Administrative Science Quarterly* 56(2): 159–201.

Davis, M. 2006. *Classic Chic: Music, Fashion and Modernism.* Berkeley: University of California Press.

Davis, M. H. 1983. "Measuring individual differences in empathy: Evidence for a multidimensional approach." *Journal of Personality and Social Psychology* 44(1): 113.

———. 1994. *Empathy: A Social Psychological Approach.* Boulder, Colo.: Westview Press.

de Tocqueville, A. 1966. *Democracy in America.* Edited by J. P. Mayer and Max Lerner. Translated by George Lawrence. New York: Harper & Row.

———. 2004. *Democracy in America.* Translated by A. Goldhammer. New York: Library of America.

Dewey, J. 1916. "What pragmatism means by practical." In *Essays in Experimental Logic.* Chicago: University of Chicago Press.

———. 1930. *Human Nature and Conduct, an Introduction to Social Psychology.* New York: Modern Library.

———. (1922) 2002. *Human Nature and Conduct.* North Chelmsford, Mass.: Courier.

Dewey, J., and A. F. Bentley. 1949. *Knowing and the Known.* Boston: Beacon Press.

DiMaggio, P. 1992. "Cultural boundaries and structural change: The extension of the high-culture model to theatre, opera, and the dance, 1900–1940." In *Cultivating Differences: Symbolic Boundaries and the Making of Inequality.* Edited by Michèle Lamont and Marcel Fournier. Chicago: University of Chicago Press.

Dougherty, D. 1992a. "Interpretive barriers to successful product innovation in large firms." *Organization Science* 3(2): 179–202.

———. 1992b. "A practice centered model of organizational renewal through product innovation." *Strategic Management Journal* 13(S1): 77–92.

———. 2002. "Grounded theory research methods." In *Companion to Organizations.* Edited by J. Baum. Oxford, England: Blackwell.

———. 2004. "Organizing practices in services: Capturing practice-based knowledge for innovation." *Strategic Organization* 2(1): 35–64.

———. 2006. "Organizing for innovation in the 21st century." In *The Sage Handbook of Organization Studies: Second Edition*, 598–617. Edited by S. R. Clegg, C. Hardy, T. Lawrence, and W. R. Nord. Thousand Oaks, Calif.: Sage.

Dougherty, D., and C. Hardy. 1996. "Sustained product innovation in large, mature organizations: Overcoming innovation-to-organization problems." *Academy of Management Journal* 39(5): 1120–1153.

Dougherty, D., and T. Heller. 1994. "The illegitimacy of successful product innovation in established firms." *Organization Science* 5(2): 200–218.

Duncan, W. R. 1996. *A Guide to the Project Management Body of Knowledge*. Newton Square, Pa.: Project Management Institute.

Edmondson, A. C., R. M. Bohmer, and G. P. Pisano. 2001. "Disrupted routines: Team learning and new technology implementations in hospitals." *Administrative Science Quarterly* 46: 685–716.

Edmondson, A. C., and I. M. Nembhard. 2009. "Product development and learning in project teams: The challenges are the benefits." *Journal of Product Innovation Management* 26(2): 123–138.

Eisenhardt, K. M., and J. A. Martin. 2000. "Dynamic capabilities: What are they?" *Strategic Management Journal* 21(10–11): 1105–1121.

Emirbayer, M., and A. Mische. 1998. "What is agency?" *American Journal of Sociology* 103(4): 962–1023.

Ewenstein, B., and J. Whyte. 2009. "Knowledge practices in design: The role of visual representations as epistemic objects." *Organization Studies* 30(1): 7–30.

Faraj, S., and Y. Xiao. 2006. "Coordination in fast-response organizations." *Management Science* 52(8): 1155–1169.

Farjoun, M. 2010. "Beyond dualism: Stability and change as a duality." *Academy of Management Review* 35(2): 202–225.

Feld, S. L. 1981. "The focused organization of social ties." *American Journal of Sociology* 86(5): 1015–1035.

———. 1982. "Social structural determinants of similarity among associates." *American Sociological Review* 47(6): 797–801.

Feldman, M. S. 2000. "Organizational routines as a source of continuous change." *Organization Science* 11(6): 611–629.

Feldman, M.S., and B. T. Pentland. 2003. "Reconceptualizing organizational routines as a source of flexibility and change." *Administrative Science Quarterly* 48(1): 94–118.

Felin, T., and N. J. Foss. 2005. "Strategic organization: A field in search of micro-foundations." *Strategic Organization* 3(4): 441.

Fermigier, A. 1967. "Jean Cocteau et Paris 1920." *Annales: Economies, Societes, Civilisations* 22(3): 495–513.

Fernandez, R. M., and R. V. Gould. 1994. "A dilemma of state power: Brokerage and influence in the national health policy domain." *American Journal of Sociology* 99(6): 1455–1491.

Fine, S. 1969. *Sit-down: The General Motors Strike of 1936–1937*. Ann Arbor: University of Michigan Press.

Fleming, L., S. Mingo, and D. Chen. 2007. "Collaborative brokerage, generative creativity, and creative success." *Administrative Science Quarterly* 52(3): 443–475.

Fleming, L., and O. Sorenson. 2004. "Science as a map in technological search." *Strategic Management Journal* 25(8–9): 909–928.

Fligstein, N. 1997. "Social skill and institutional theory." *American Behavioral Scientist* 40(4): 397–405.

———. 2001. "Social skill and the theory of fields." *Social Theory* 19: 105–125.

Fligstein, N., and D. McAdam. 2011. "Toward a general theory of strategic action fields." *Sociological Theory* 29(1): 1–26.

———. 2012. *A Theory of Fields*. Oxford: Oxford University Press.

Foss, N. J. 2003. "Selective intervention and internal hybrids: Interpreting and learning from the rise and decline of the Oticon spaghetti organization." *Organization Science* 14(3): 331–349.

Fox-Wolfgramm, S. J., K. B. Boal, and J. G. Hunt. 1998. "Organizational adaptation to institutional change: A comparative study of first-order change in prospector and defender banks." *Administrative Science Quarterly* 43(1): 87–127.

Freeman, L. C. 1977. "A set of measures of centrality based on betweenness." *Sociometry* 40(1): 35–41.

———. 1979. "Centrality in social networks conceptual clarification." *Social Networks* 1(3): 215–239.

Freeman, L., and A. Romney. 1987. "Words, deeds and social structure: A preliminary study of the reliability of informants." *Human Organization* 46(4): 330–334.

Friedman, T. L. 2005. *The World Is Flat: A Brief History of the Twenty-first Century*. New York: Farrar, Straus and Giroux.

Galinsky, A. D., W. W. Maddux, D. Gilin, and J. B. White. 2008. "Why it pays to get inside the head of your opponent: The differential effects of perspective taking and empathy in negotiations." *Psychological Science* 19(4): 378–384.

Gamson, W. A. 1975. *The Strategy of Social Protest.* Homewood, Ill.: Dorsey Press.

Gamson, W. A., B. Fireman, and S. Rytina. 1982. *Encounters with Unjust Authority.* Homewood, Ill.: Dorsey Press.

Gangestad, S. W., and M. Snyder. 2000. "Self-monitoring: Appraisal and reappraisal." *Psychological Bulletin* 126(4): 530.

Gann, D. M., and A. J. Salter. 2000. "Innovation in project-based, service-enhanced firms: The construction of complex products and systems." *Research Policy* 29(7): 955–972.

Garafola, L. 1989. *Diaghilev's Ballets Russes.* New York: Da Capo Press.

Garfinkel, H. 1963. "A conception of and experiments with 'trust' as a condition of concerted stable actions." In *The Production of Reality: Essays and Readings on Social Interaction,* 381–392. Edited by Jodi O'Brien. Thousand Oaks, Calif.: Sage.

———. 1967. *Studies in Ethnomethodology.* Englewood Cliffs, N.J.: Prentice-Hall.

Gargiulo, M., and A. Rus. 2002. "Access and mobilization: Social capital and top management response to market shocks." Fontainebleau: INSEAD.

Gartner, W. B. 1989. "Some suggestions for research on entrepreneurial traits and characteristics." *Entrepreneurship Theory and Practice* 14(1): 27–38.

Gartner, W. B., B. J. Bird, and J. A. Starr. 1992. "Acting as if: Differentiating entrepreneurial from organizational behavior." *Entrepreneurship Theory and Practice* 16(3): 13–31.

Garud, R., and P. Karnøe. 2003. "Bricolage versus breakthrough: Distributed and embedded agency in technology entrepreneurship." *Research Policy* 32(2): 277–300.

Gavetti, G. 2005. "Cognition and hierarchy: Rethinking the microfoundations of capabilities' development." *Organization Science* 16(6): 599–617.

Geertz, C. 1973. *The Interpretation of Cultures: Selected Essays.* Vol. 5019. New York: Basic Books.

Gersick, C. J., and J. R. Hackman. 1990. "Habitual routines in task-performing groups." *Organizational Behavior and Human Decision Processes* 47(1): 65–97.

Gerson, E. M., and S. L. Star. 1986. "Analyzing due process in the workplace." *ACM Transactions on Information Systems (TOIS)* 4(3): 257–270.

Ghonim, W. 2012. *Revolution 2.0: The Power of the People Is Greater than the People in Power: A Memoir.* Boston: Houghton Mifflin Harcourt.

Giannoni, G. 2013. *Howard Schultz's Secrets of Success: Ambition and Charisma with a Social Conscience.* Baquiano Books.

Gibson, C. B. 1999. "Do they do what they believe they can? Group efficacy and group effectiveness across tasks and cultures." *Academy of Management Journal* 42(2): 138–152.

Gibson, C. B., and J. Birkinshaw. 2004. "The antecedents, consequences, and mediating role of organizational ambidexterity." *Academy of Management Journal* 47(2): 209–226.

Gilbert, C. G. 2005. "Unbundling the structure of inertia: Resource versus routine rigidity." *Academy of Management Journal* 48(5): 741–763.

Girard, M., and D. Stark. 2002. "Distributing intelligence and organizing diversity in new media projects." *Sociedade e estado* 17(1): 153–192.

Giroux, H., and J. R. Taylor. 2002. "The justification of knowledge: Tracking the translations of quality." *Management Learning* 33(4): 497–517.

Gladwell, M. 2010. "Small change." *New Yorker,* October 4, 42–49.

———. 2011. "Creation myth: Xerox PARC, Apple, and the truth about innovation." *New Yorker,* May 16.

Gladwell, M., and C. Shirky. 2011. "From innovation to revolution." *Foreign Affairs* 90(2): 153–154.

Goffman, E. 1959. *The Presentation of Self in Everyday Life.* Garden City, N.Y.: Doubleday.

———. 1961. *Encounters: Two Studies in Social Interaction.* Indianapolis: Bobbs-Merrill.

———. 1974. *Frame Analysis: An Essay on the Organization of Experience.* Cambridge, Mass.: Harvard University Press.

Goldberg, A., S. B. Srivastava, V. G. Manian, W. Monroe, and C. Potts. Forthcoming. "Fitting in or standing out: The tradeoffs of structural and cultural embeddedness." *American Sociological Review* 81(6): 1190–1222.

Goldrick-Rab, S. 2006. "Following their every move: An investigation of social-class differences in college pathways." *Sociology of Education* 79(1): 67–79.

Goodwin, D. K. 2005. *Team of Rivals: The Political Genius of Abraham Lincoln.* New York: Simon & Schuster.

Gould, R. V., and R. Fernandez. 1989. "Structures of mediation: A formal approach to brokerage in transaction networks." *Sociological Methodology* 19: 89–126.

Gourlay, S. 2004. "'Tacit knowledge': The variety of meanings in empirical research." Fifth European Conference on Organizational Knowledge, Learning and Capabilities, April 2–3, Innsbruck, Austria.

Grabher, G. 2002. "Cool projects, boring institutions: Temporary collaboration in social context." *Regional Studies* 36(3): 205–214.

———. 2004. "Learning in projects, remembering in networks? Communality, sociality, and connectivity in project ecologies." *European Urban and Regional Studies* 11(2): 103–123.

Granovetter, M. S. 1973. "The strength of weak ties." *American Journal of Sociology* 78(6): 1360–1380.

———. 2005. "The impact of social structure on economic outcomes." *Journal of Economic Perspectives* 19(1): 33–50.

Grant, A. M. 2013. *Give and Take: Why Helping Others Drives Our Success.* New York: Penguin.

Grant, R. M. 1996. "Toward a knowledge based theory of the firm." *Strategic Management Journal* 17(S2): 109–122.

Grigoriev, S. L. 1953. *The Diaghilev Ballet, 1909–1929.* Translated and edited by Vera Bowen. Hammondsworth: Penguin.

Grosser, T. J., D. Obstfeld, G. Labianca, and S. P. Borgatti. 2015. "Brokerage orientations, social networks, and innovation support: Validating the discrete brokerage orientation scale." Working paper, University of Connecticut School of Business, Storrs.

Gulati, R., and S. B. Srivastava. 2014. "Bringing agency back into network research: Constrained agency and network action." In *Research in the Sociology of Organizations: Contemporary Perspectives on Organizational Social Networks.* Vol. 40. Edited by D. J. Brass, G. Labianca, A. Mehra, D. S. Halgin, and S. P. Borgatti. Bingley: Emerald Group Publishing.

Gupta, A. K., K. G. Smith, and C. E. Shalley. 2006. "The interplay between exploration and exploitation." *Academy of Management Journal* 49(4): 693–706.

Hagel, J., J. Seely Brown, and L. Davison. 2010. *The Power of Pull.* New York: Basic Books.

Hallen, B. L. 2008. "The causes and consequences of the initial network positions of new organizations: From whom do entrepreneurs receive investments?" *Administrative Science Quarterly* 53(4): 685–718.

Hansen, M. T. 1999. "The search-transfer problem: The role of weak ties in sharing knowledge across organization subunits." *Administrative Science Quarterly* 44(1): 82–111.

Hargadon, A. B. 2002. "Brokering knowledge: Linking learning and innovation." *Research in Organizational Behavior* 24: 41–85.

———. 2003. *How Breakthroughs Happen: The Surprising Truth about How Companies Innovate.* Cambridge, Mass.: Harvard Business Press.

Hargadon, A. B., and R. I. Sutton. 1997. "Technology brokering and innovation in a product development firm." *Administrative Science Quarterly* 42(4): 716–749.

Hasher, L., and R. T. Zacks. 1984. "Automatic processing of fundamental information: The case of frequency of occurrence." *American Psychologist* 39(12): 1372.

Hayek, F. A. 1948. *Individualism and Economic Order.* Chicago: University of Chicago Press.

Hedström, P., and R. Swedberg. 1998. *Social Mechanisms: An Analytical Approach to Social Theory.* Cambridge: Cambridge University Press.

Henderson, K. 1998. *On Line and On Paper: Visual Representations, Visual Culture, and Computer Graphics in Design Engineering.* Cambridge, Mass.: MIT Press.

Heritage, J. 1984. *Garfinkel and Ethnomethodology.* Cambridge: Polity Press.

Higgins, T. 1998. "Promotion and prevention: Regulatory focus as a motivational principle." In *Advances in Experimental Social Psychology,* no. 30, 1–46. Edited by M. P. Zanna. San Diego: Academic Press.

Hobday, M. 2000. "The project-based organisation: An ideal form for managing complex products and systems?" *Research Policy* 29(7): 871–893.

Howard-Grenville, J. A. 2005. "The persistence of flexible organizational routines: The role of agency and organizational context." *Organization Science* 16(6): 618–636.

Huesca, R. 2001. *Triomphes et scandale: La Belle Epoque de Ballets Russes.* Paris: Hermann.

Ibarra, H. 1989. "Centrality and innovativeness: Effects of social network position on innovation involvement." Dissertation, Yale University.

———. 1993. "Network centrality, power, and innovation involvement: Determinants of technical and administrative roles." *Academy of Management Journal* 36(3): 471–501.

Inkpen, A. C., and A. Dinur. 1998. "Knowledge management processes and international joint ventures." *Organization Science* 9(4): 454–468.

James, W. 1975. *Pragmatism.* Vol. 1. Cambridge, Mass.: Harvard University Press.

Jasper, J. M. 1997. *The Art of Moral Protest.* Chicago: University of Chicago Press.

———. 2004. "Intellectual cycles of social movement research." In *Self, Social Structure, and Beliefs: Explorations in Sociology*, 234–253. Edited by J. C. Alexander, G. T. Marx, and C. L. Williams. Berkeley: University of California Press.

Jelinek, M., and C. Schoonhoven. 1993. *The Innovation Marathon: Lessons from High Technology Companies.* San Francisco: Jossey-Bass.

Jick, T. D. 1979. "Mixing qualitative and quantitative methods: Triangulation in action." *Administrative Science Quarterly* 24(4): 602–611.

Joas, H. 1996. *The Creativity of Action.* Chicago: University of Chicago Press.

———. 1997. *GH Mead: A Contemporary Re-examination of His Thought.* Cambridge, Mass.: MIT Press.

Kacperczyk, A., J. P. Davis, and O. Hahl. 2011. "Knowledge asymmetry in brokerage: Secret network sources of broker's position and power." Working paper, MIT Sloan School of Management.

Kaplan, S. 2008. "Framing contests: Strategy making under uncertainty." *Organization Science* 19(5): 729–752.

Kaplan, S., J. Milde, and R. S. Cowan. Forthcoming. "Symbiotic practices in boundary spanning: Bridging the cognitive and political divides in interdisciplinary research." *Academy of Management Journal.*

Kellogg, K. C. 2014. "Brokerage professions and implementing reform in an age of experts." *American Sociological Review* 79(5): 912–941.

Kellogg, K. C., W. J. Orlikowski, and J. Yates. 2006. "Life in the trading zone: Structuring coordination across boundaries in postbureaucratic organizations." *Organization Science* 17(1): 22–44.

Kirzner, I. M. 1989. *Discovery, Capitalism, and Distributive Justice.* Oxford: Basil Blackwell.

Klein, K. J., J. C. Ziegert, A. P. Knight, and Y. Xiao. 2006. "Dynamic delegation: Shared, hierarchical, and deindividualized leadership in extreme action teams." *Administrative Science Quarterly* 51(4): 590–621.

Kogut, B. 2000. "The network as knowledge: Generative rules and the emergence of structure." *Strategic Management Journal* 21(3): 405–425.

Kogut, B., and U. Zander. 1992. "Knowledge of the firm, combinative capabilities, and the replication of technology." *Organization Science* 3(3): 383–397.

———. 1996. "What firms do? Coordination, identity, and learning." *Organization Science* 7(5): 502–518.

Krackhardt, D. 1990. "Assessing the political landscape: Structure, cognition, and power in organizations." *Administrative Science Quarterly* 35: 342–369.

———. 1999. "The ties that torture: Simmelian tie analysis in organizations." *Research in the Sociology of Organizations* 16(1): 183–210.

Kraus, H. 1947. *The Many and the Few.* Los Angeles: Plantin.

Kurlaender, M., and S. M. Flores. 2005. "The racial transformation of higher education." In *Higher Education and the Color Line: College Access, Racial Equity, and Social Change*, 11–32. Edited by G. Orfield, P. Marin, and C. Horn. Cambridge, Mass.: Harvard Education Press.

Lampel, J. 2001. "The core competencies of effective project execution: The challenge of diversity." *International Journal of Project Management* 19(8): 471–483.

Latour, B. 1986. "Visualization and cognition." *Knowledge and Society* 6: 1–40.

———. 1987. *Science in Action: How to Follow Scientists and Engineers through Society.* Cambridge, Mass.: Harvard University Press.

———. 2005. *Reassembling the Social: An Introduction to Actor-Network-Theory.* Oxford: Oxford University Press.

Lave, J. 1991. "Situating learning in communities of practice." *Perspectives on Socially Shared Cognition* 2: 63–82.

Lave, J., and E. Wenger. 1991. *Communities of Practice.* Cambridge: Cambridge University Press.

Lee, J.Y., D. G. Bachrach, and K. Lewis. 2014. "Social network ties, transactive memory, and performance in groups." *Organization Science* 25(3): 951–967.

Levine, J. M., E. T. Higgins, and H.-S. Choi. 2000. "Development of strategic orientations in groups." *Organizational Behavior and Human Decision Processes* 82: 88–101.

Levitt, B., and J. G. March. 1988. "Organizational learning." *Annual Review of Sociology* 14: 319–340.

Li, M., and T.Y. Choi. 2009. "Triads in services outsourcing: Bridge, bridge decay and bridge transfer." *Journal of Supply Chain Management* 45(3): 27–39.

Lillrank, P. 2003. "The quality of standard, routine and nonroutine processes." *Organization Studies* 24(2): 215–233.

Locke, K. 1996. "Rewriting the discovery of grounded theory after 25 years?" *Journal of Management Inquiry* 5: 239–245.

Locke, K. D., and K. Golden-Biddle. 1997. *Composing Qualitative Research*. Thousand Oaks, Calif.: Sage.

Lodge, D. 2012. *The Art of Fiction*. New York: Random House.

Long Lingo, E. L., and S. O'Mahony. 2010. "Nexus work: Brokerage on creative projects." *Administrative Science Quarterly* 55(1): 47–81.

Lorrain, F., and H. C. White. 1971. "Structural equivalence of individuals in social networks." *Journal of Mathematical Sociology* 1(1): 49–80.

Louis, M. R., and R. I. Sutton. 1991. "Switching cognitive gears: From habits of mind to active thinking." *Human Relations* 44(1): 55–76.

Luhmann, N. 1995. *Social Systems*. Stanford, Calif.: Stanford University Press.

Lundin, R. A., and A. Söderholm. 1995. "A theory of the temporary organization." *Scandinavian Journal of Management* 11(4): 437–455.

Mailloux, S. 2011. "Euro-American rhetorical pragmatism: Democratic deliberation, humanist controversies, and purposeful mediation." *Pragmatism Today* 2(2): 81–91.

Majchrzak, A., P. H. More, and S. Faraj. 2012. "Transcending knowledge differences in cross-functional teams." *Organization Science* 23(4): 951–970.

Malone, T. W., K. Crowston, J. Lee, B. Pentland, C. Dellarocas, G. Wyner, and E. O'Donnell. 1999. "Tools for inventing organizations: Toward a handbook of organizational processes." *Management Science* 45(3): 425–443.

March, J. G. 1971. "The technology of foolishness." *Civiløkonomen* (Copenhagen) 18(4): 4–12.

———. 1991. "Exploration and exploitation in organizational learning." *Organization Science* 2(1): 71–87.

March, J. G., and H. A. Simon. 1958. *Organizations*. New York: Wiley.

Marsden, P.V. 1982. "Brokerage behavior in restricted exchange networks." In *Social Structure and Network Analysis*, 201–218. Edited by P.V. Marsden and N. Lin. Beverly Hills, Calif.: Sage.

Massey, D. 2005. *For Space*. London: Sage.

Mayer, R. C., J. H. Davis, and F. D. Schoorman. 1995. "An integrative model of organizational trust." *Academy of Management Review* 20(3): 709–734.

McAdam, D. 1982. *Political Process and the Development of Black Insurgency 1930–1970*. Chicago: University of Chicago Press.

———. 1990. *Freedom Summer*. Oxford: Oxford University Press.

———. 2003. "Beyond structural analysis: Toward a more dynamic understanding of social movements." In *Social Movements and Networks: Relational Approaches to Collective Action*. Edited by M. Diani and D. McAdam. Oxford: Oxford University Press.

McAdam, D., J. D. McCarthy, and M. Zald. 1988. "Social movements." In *Handbook of Sociology*. Edited by N. Smelser. Newbury Park, Calif.: Sage.

———. 1996. "Opportunities, mobilizing structures, and framing processes: Toward a synthetic, comparative perspective on social movements." In *Comparative Perspectives on Social Movements: Political Opportunities, Mobilizing Structures, and Cultural Framings*. Edited by D. McAdam, J. D. McCarthy, and M. N. Zald. New York: Cambridge University Press.

McAdam, D., and R. Paulsen. 1993. "Specifying the relationship between social ties and activism." *American Journal of Sociology* 99(3): 640–667.

McAdam, D., S. Tarrow, and C. Tilly. 2001. *Dynamics of Contention*. Cambridge: Cambridge University Press.

McCloskey, D. N. 1998. *The Rhetoric of Economics*. Madison: University of Wisconsin Press.

McFarland, D. A. 2004. "Resistance as a social drama: A study of change oriented encounters." *American Journal of Sociology* 109(6): 1249–1318.

McMullen, J. S., and D. A. Shepherd. 2006. "Entrepreneurial action and the role of uncertainty in the theory of the entrepreneur." *Academy of Management Review* 31(1): 132–152.

Mead, G. H. 1934. *Mind, Self and Society: From the Standpoint of a Social Behaviourist.* Chicago: University of Chicago Press.

———.1938. *The Philosophy of the Act.* Chicago: University of Chicago Press.

Mehra, A., M. Kilduff, and D. J. Brass. 2001. "The social networks of high and low self-monitors: Implications for workplace performance." *Administrative Science Quarterly* 46(1): 121–146.

Mills, C. W. 1939. "Language, logic, and culture." *American Sociological Review* 4(5): 670–680.

Miner, A. S. 1991. "Organizational evolution and the social ecology of jobs." *American Sociological Review* 56: 772–785.

Miner, A. S., P. Bassof, and C. Moorman. 2001. "Organizational improvisation and learning: A field study." *Administrative Science Quarterly* 46(2): 304–337.

Miner, A. S., and E. S. Estler. 1985. "Accrual mobility: Job mobility in higher education through responsibility accrual." *Journal of Higher Education* 56: 121–143.

Mintzberg, H. 1993. *Structure in Fives: Designing Effective Organizations.* Englewood Cliffs, N.J.: Prentice-Hall.

Mintzberg, H., D. Raisinghani, and A. Theoret. 1976. "The structure of 'unstructured' decision processes." *Administrative Science Quarterly* 21(2): 246–275.

Mische, A., and H. White. 1998. "Between conversation and situation: Public switching dynamics across network domains." *Social Research* 65(3): 695–724.

Moldoveanu, M., and J. Baum. 2014. *Epinets: The Epistemic Structure and Dynamics of Social Networks.* Stanford, Calif.: Stanford University Press.

Munro, T. 1951. "'The Afternoon of a Faun' and the interrelation of the arts." *Journal of Aesthetics and Art Criticism* 10(2): 95–111.

Nahapiet, J., and S. Ghoshal. 1998. "Social capital, intellectual capital, and the organizational advantage." *Academy of Management Review* 23(2): 242–266.

Nectoux, J. M. 1992. "Shéhérazade, danse, musiques." *Romantisme* 22(78): 35–42.

Nelson, R., and S. Winter. 1982. *An Evolutionary Theory of Economic Change.* Cambridge, Mass.: Belknap.

Nonaka, I., and H. Takeuchi. 1995. *The Knowledge-creating Company: How Japanese Companies Create the Dynamics of Innovation.* New York: Oxford University Press.

Obstfeld, D. 2001. "Telling more of what we know: Examining the social processes of knowledge creation and innovation." Dissertation, University of Michigan, Ann Arbor.

———. 2005. "Social networks, the *tertius iungens* orientation, and involvement in innovation." *Administrative Science Quarterly* 50(1): 100–130.

———. 2012. "Creative projects: A less routine approach toward getting new things done." *Organization Science* 23(6): 1571–1592.

Obstfeld, D., S. P. Borgatti, and J. P. Davis. 2014. "Brokerage as a process: Decoupling third party action from social network structure." *Contemporary Perspectives on Organizational Social Networks* 40: 135–159.

Okhuysen, G. A., and B. A. Bechky. 2009. "Coordination in organizations: An integrative perspective." *Academy of Management Annals* 3(1): 463–502.

Oliver, P. E. 2003. "Mechanisms of contention." *Mobilization* 8(1): 119–121.

O'Mahony, S., and F. Ferraro. 2007. "The emergence of governance in an open source community." *Academy of Management Journal* 50(5): 1079–1106.

Orlikowski, W. J. 2000. "Using technology and constituting structures: A practice lens for studying technology in organizations." *Organization Science* 11(4): 404–428.

———. 2002. "Knowing in practice: Enacting a collective capability in distributed organizing." *Organization Science* 13(3): 249–273.

Orr, J. E. 1996. *Talking about Machines: An Ethnography of a Modern Job.* Ithaca, N.Y.: Cornell University Press.

Ozcan, P., and K. M. Eisenhardt. 2009. "Origin of alliance portfolios: Entrepreneurs, network strategies, and firm performance." *Academy of Management Journal* 52(2): 246–279.

Ozick, C. 1989. *Metaphor and Memory.* New York: Knopf.

Padgett, J. F., and C. K. Ansell. 1993. "Robust action and the rise of the Medici, 1400–1434." *American Journal of Sociology* 98(6): 1259–1319.

Padgett, J. F., and W. W. Powell. 2012. *The Emergence of Organizations and Markets*. Princeton, N.J.: Princeton University Press.

Parker, I. 2015. "The shape of things to come: How an industrial designer became Apple's greatest product." *New Yorker*, February 23.

Parsons, T. 1951. *The Social System*. New York and London: Free Press and Collier Macmillan.

———. 1966. "Societies." In *Evolutionary and Comparative Perspectives*. Englewood Cliffs, N.J.: Prentice-Hall.

Pava, C. 1986. "Redesigning sociotechnical systems design: Concepts and methods for the 1990s." *Journal of Applied Behavioral Science* 22(3): 201–221.

Pentland, B. T. 2004. "Towards an ecology of inter-organizational routines: A conceptual framework for the analysis of net-enabled organizations." In *System Sciences, 2004: Proceedings of the 37th Annual Hawaii International Conference on System Sciences*, 264–271. IEEE.

Pentland, B. T., and M. S. Feldman. 2005. "Organizational routines as a unit of analysis." *Industrial and Corporate Change* 14(5): 793–815.

Pentland, B. T., and H. H. Rueter. 1994. "Organizational routines as grammars of action." *Administrative Science Quarterly* 39: 484–510.

Perrow, C. 1967. "A framework for the comparative analysis of organizations." *American Sociological Review* 32: 194–208.

Pfeffer, F. T., and F. Hertel. 2014. "How has educational expansion shaped social mobility trends in the United States?" *PSC Research Report*, no. 14–817.

Pfeffer, J. 2010. "Power play." *Harvard Business Review* 88(7–8): 84–92.

Piketty, T. 2014. *Capital in the 21st Century*. Cambridge, Mass.: Harvard University Press.

Podolny, J. M. 2001. "Networks as the pipes and prisms of the market." *American Journal of Sociology* 107(1): 33–60.

Podolny, J. M., and J. N. Baron. 1997. "Resources and relationships: Social networks and mobility in the workplace." *American Sociological Review* 62: 673–693.

Polanyi, M. 1958. *Personal Knowledge: Towards a Post-Critical Philosophy*. Chicago: University of Chicago Press.

———. 1966. "The logic of tacit inference." *Philosophy* 41(155): 1–18.

Pollock, T. G., J. F. Porac, and J. B. Wade. 2004. "Constructing deal networks: Brokers as network 'architects' in the US IPO market and other examples." *Academy of Management Review* 29(1): 50–72.

Prencipe, A., and F. Tell. 2001. "Inter-project learning: Processes and outcomes of knowledge codification in project-based firms." *Research Policy* 30(9): 1373–1394.

Quintane, E., and G. Carnabuci. 2016. "How do brokers broker? Tertius Gaudens, Tertius Iungens, and the temporality of structural holes." *Organization Science* 27(6): 1343–1360.

Raisch, S., and J. Birkinshaw. 2008. "Organizational ambidexterity: Antecedents, outcomes, and moderators." *Journal of Management* 34(3): 375–409.

Reagans, R., and B. McEvily. 2003. "Network structure and knowledge transfer: The effects of cohesion and range." *Administrative Science Quarterly* 48(2): 240–267.

———. 2008. "Contradictory or compatible? Reconsidering the 'trade-off' between brokerage and closure on knowledge sharing." *Advances in Strategic Management* 25: 275–313.

Reagans, R., and E. W. Zuckerman. 2001. "Networks, diversity, and productivity: The social capital of corporate R&D teams." *Organization Science* 12(4): 502–517.

Reagans, R., E. Zuckerman, and B. McEvily. 2004. "How to make the team: Social networks vs. demography as criteria for designing effective teams." *Administrative Science Quarterly* 49(1): 101–133.

Reger, R. K., and T. B. Palmer. 1996. "Managerial categorization of competitors: Using old maps to navigate new environments." *Organization Science* 7(1): 22–39.

Rice, P., and P. Waugh. 2001. *Modern Literary Theory: A Reader*. Oxford: Oxford University Press.

Ries, E. 2011. *The Lean Startup: How Today's Entrepreneurs Use Continuous Innovation to Create Radically Successful Businesses*. New York: Random House.

Rodan, S., and D. C. Galunic. 2004. "More than network structure: How knowledge heterogeneity influences managerial performance and innovativeness." *Strategic Management Journal* 25: 541–556.

Rogers, E. M. 1983. *Diffusion of Innovations*. New York: Collier Macmillan.

———. 2003. *Diffusion of Innovations*. New York: Simon & Schuster.

Rosenkopf, L., and M. L. Tushman. 1994. "The coevolution of technology and organization." In *Evolutionary Dynamics of Organizations*, 403–424. Edited by J. Baum and J. Singh. New York: Oxford University Press.

Ruef, M. 2009. "Entrepreneurial groups." In *Historical Foundations of Entrepreneurship Research*, 205–228. Edited by H. Landström and F. Lohrke. Cheltenham: Edward Elgar Press.

———. 2010. *The Entrepreneurial Group: Social Identities, Relations, and Collective Action*. Princeton, N.J.: Princeton University Press.

Ryall, M. D., and O. Sorenson. 2007. "Brokers and competitive advantage." *Management Science* 53(4): 566–583.

Salovey, P., and J. D. Mayer. 1990. "Emotional intelligence." *Imagination, Cognition and Personality* 9(3): 185–211.

Sarasvathy, S. D. 2001. "Causation and effectuation: Toward a theoretical shift from economic inevitability to entrepreneurial contingency." *Academy of Management Review* 26(2): 243–263.

———. 2008. *Effectuation: Elements of Entrepreneurial Orientation*. Cheltenham: Edward Elgar Press.

Schmickl, C., and A. Kieser. 2008. "How much do specialists have to learn from each other when they jointly develop radical product innovations?" *Research Policy* 37(3): 473–491.

Schumpeter, J. A. 1926. *Theorie der wirtschaftlichen Entwicklung: Eine Untersuchung über Unternehmergewinn, Kapital, Kredit, Zins und den Konjunkturzyklus*. Munich: Duncker & Humblot.

———. 1934. *Theory of Economic Development*. Cambridge, Mass.: Harvard University Press.

———. 1947. "The creative response in economic history." *Journal of Economic History* 7(2): 149–159.

Schütz, A. 1944. "The stranger: An essay in social psychology." *American Journal of Sociology* 49(6): 499–507.

———. 1967. *The Phenomenology of the Social World*. Evanston, Ill.: Northwestern University Press.

Sgourev, S. V. 2015. "Brokerage as catalysis: How Diaghilev's *Ballets Russes* escalated modernism." *Organization Studies* 36(3): 343–361.

Shane, S. 2012. "Reflections on the 2010 AMR decade award: Delivering on the promise of entrepreneurship as a field of research." *Academy of Management Review* 37(1): 10–20.

Shane, S., and S. Venkataraman. 2000. "The promise of entrepreneurship as a field of research." *Academy of Management Review* 25(1): 217–226.

Shi, W., L. Markoczy, and G. C. Dess. 2009. "The role of middle management in the strategy process: Group affiliation, structural holes, and tertius iungens." *Journal of Management* 35(6): 1453–1480.

Shirky, C. 2008. *Here Comes Everybody: The Power of Organizing without Organizations*. New York: Penguin.

Shklovsky, V. 1988. "Art as technique." In *Modern Criticism and Theory: A Reader*. Edited by D. Lodge. New York: Longman.

———. 1990. *Theory of Prose*. Translated by B. Sher. Elmwood Park, Ill.: Dalkey Archive Press.

Simmel, G., and K. H. Wolff. 1950. *The Sociology of Georg Simmel*. Vol. 92892. New York: Simon & Schuster.

Simon, H. A. 1945. *Administrative Behavior*. London: Macmillan.

Sitkin, S. B., K. E. See, C. C. Miller, M. W. Lawless, and A. M. Carton. 2011. "The paradox of stretch goals: Organizations in pursuit of the seemingly impossible." *Academy of Management Review* 36(3): 544–566.

Small, M. L. 2009. *Unanticipated Gains: Origins of Network Inequality in Everyday Life*. New York: Oxford University Press.

Smith, J. Z. 1985. "Double play." In *Engaging the Humanities at the University of Chicago*, 57–60. Edited by P. Desan. Chicago: University of Chicago Press.

Smith, M. F. 1996. "Issue status and social movement organization maintenance: Two case studies in rhetorical diversification." Dissertation, Purdue University.

Smith, W. K., and M. L. Tushman. 2005. "Managing strategic contradictions: A top management model for managing innovation streams." *Organization Science* 16(5): 522–536.

Snow, D. A. 2006. "Are there really awkward movements or only awkward research relationships?" *Mobilization: An International Quarterly* 11(4): 495–498.

Snow, D., D. Cress, L. Downey, and A. Jones. 1998. "Disrupting the 'quotidian': Reconceptualizing the relationship between breakdown and the emergence of collective action." *Mobilization: An International Quarterly* 3(1): 1–22.

Snow, D. A., E. B. Rochford Jr., S. K. Worden, and R. D. Benford. 1986. "Frame alignment processes, micromobilization, and movement participation." *American Sociological Review* 51: 464–481.

Sobek, D. K. 1997. "Principles that shape product development systems: A Toyota-Chrysler comparison." Dissertation, University of Michigan, Ann Arbor.

Soda, M. Tortoriello, and A. Ioriao. 2015. "It is how you broker: The contingent effects of individuals' behavioral orientations on the relationship between structural holes and performance." Working paper, Bocconi University, Milan.

Sorenson, O., and T. E. Stuart. 2001. "Syndication networks and the spatial distribution of venture capital investments." *American Journal of Sociology* 106(6): 1546–1588.

Spender, J. C. 1996. "Making knowledge the basis of a dynamic theory of the firm." *Strategic Management Journal* 17(S2): 45–62.

———. 2005. "Organizations as knowledge systems: Knowledge, learning and dynamic capabilities." *Organization Studies* 26(1): 137–143.

Spinosa, C., F. Flores, and H. L. Dreyfus. 1997. *Disclosing New Worlds: Entrepreneurship, Democratic Action, and the Cultivation of Solidarity.* Cambridge, Mass.: MIT Press.

Spiro, E. S., R. M. Acton, and C. T. Butts. 2013. "Extended structures of mediation: Re-examining brokerage in dynamic networks." *Social Networks* 35(1): 130–143.

Srivastava, S. B. 2015. "Intraorganizational network dynamics in times of ambiguity." *Organization Science* 26: 1365–1380.

Srivastava, S. B., A. Goldberg, V. Govind Manian, and C. Potts. 2017. "Enculturation trajectories: Language, cultural adaptation, and individual outcomes in organizations." *Management Science.* Published online in Articles in Advance, March 2.

Star, S. L., and J. R. Griesemer. 1989. "Institutional ecology, translations and boundary objects: Amateurs and professionals in Berkeley's Museum of Vertebrate Zoology, 1907–39." *Social Studies of Science* 19(3): 387–420.

Star, S. L., and A. Strauss. 1999. "Layers of silence, arenas of voice: The ecology of visible and invisible work." *Computer Supported Cooperative Work (CSCW)* 8(1–2): 9–30.

Stene, E. O. 1940. "An approach to a science of administration." *American Political Science Review* 34(6): 1124–1137.

Stigliani, I., and D. Ravasi. 2012. "Organizing thoughts and connecting brains: Material practices and the transition from individual to group-level prospective sensemaking." *Academy of Management Journal* 55(5): 1232–1259.

Stiglitz, J. 2012. *The Price of Inequality.* New York: Norton.

Stovel, K., and L. Shaw. 2012. "Brokerage." *Annual Review of Sociology* 38: 139–158.

Strang, D., and D. Jung. 2005. "Organizational change as an orchestrated social movement: Recruitment to a 'quality initiative.'" In *Social Movements and Organization Theory*, 280–309. Edited by G. F. Davis, D. McAdam, W. R. Scott, and M. N. Zald. Cambridge: Cambridge University Press.

Strauss, A. 1988. "The articulation of project work: An organizational process." *Sociological Quarterly* 29: 163–178.

———. 1993. *Continual Permutations of Action.* New York: Aldine de Gruyter.

Strauss, A., and J. Corbin. 1990. *Basics of Qualitative Research: Grounded Theory Procedures and Techniques.* Thousand Oaks, Calif.: Sage.

Strauss, A. L., S. Fagerhaugh, B. Suczek, and W. Wiener. 1985. *Social Organization of Medical Work.* Chicago: University of Chicago Press.

Strike, V., and C. Rerup. 2016. "Mediated Sensemaking." *Academy of Management Journal* 59(3): 880–905.

Suchman, L. 1987. *Plans and Situated Actions: The Problem of Human-Machine Communication.* Cambridge: Cambridge University Press.

———. 1996. "Supporting articulation work." In *Computerization and Controversy: Value Conflicts and Social Choices*, 407–423. Edited by R. Kling. San Diego: Academic Press.

———. 2003. "Writing and reading: A response to comments on plans and situated actions." *Journal of the Learning Sciences* 12(2): 299–306.

Suchman, M. C. 1995. "Managing legitimacy: Strategic and institutional approaches." *Academy of Management Review* 20(3): 571–610.

Swedberg, R. 2007. "Rebuilding Schumpeter's theory of entrepreneurship." In *Marshall and Schumpeter on Evolution: Economic Sociology of Capitalist Development.* Edited by S. Yu and T. Nishizawa. Cheltenham: Edward Elgar Press.

Swidler, A. 1986. "Culture in action: Symbols and strategies." *American Sociological Review* 51: 273–286.

Taleb, N. 2007. *The Black Swan: The Impact of the Highly Improbable.* New York: Random House.

Tarrow, S. 2003. "Confessions of a recovering structuralist." *Mobilization* 8(1): 134–141.

Tarrow, S., and C. Tilly. 2007. "Contentious politics and social movements." In *The Oxford Handbook of Comparative Politics*, 435–460. Edited by Carles Boix and Susan C. Stokes. Oxford: Oxford University Press.

Taylor, J. R., and E. Van Every. 2000. *The Emergent Organization: Communication as Its Site and Surface.* Mahwah, N.J.: Lawrence Erlbaum Associates.

Teece, D. J. 1998. "Capturing value from knowledge assets: The new economy, markets for know-how, and intangible assets." *California Management Review* 40: 55–79.

———. 2012. "Dynamic capabilities: Routines versus entrepreneurial action." *Journal of Management Studies* 49(8): 1395–1401.

———. 2014. "A dynamic capabilities-based entrepreneurial theory of the multinational enterprise." *Journal of International Business Studies* 45(1): 8–37.

Teece, D., and G. Pisano. 1994. "The dynamic capabilities of firms: An introduction." *Industrial and Corporate Change* 3(3): 537–556.

Ter Wal, A. L., O. Alexy, J. Block, and P. G. Sandner. 2016. "The best of both worlds: The benefits of open-specialized and closed-diverse syndication networks for new ventures' success." *Administrative Science Quarterly* 61(3): 393–432.

Tilly, C. 1976. "Major forms of collective action in western Europe 1500–1975." *Theory and Society* 3(3): 365–375.

———. 1977. "Getting it together in Burgundy, 1675–1975." *Theory and Society* 4(4): 479–504.

———. 1978. *From Mobilization to Revolution.* New York: McGraw-Hill.

———. 1993. "Contentious repertoires in Great Britain, 1758–1834." *Social Science History* 17(2): 253–280.

———. 1995. "To explain political processes." *American Journal of Sociology* 100(6): 1594–1610.

———. 1997. "Means and ends of comparison in macrosociology." *Comparative Social Research* 16: 43–54.

Tilly, C., and S. Tarrow. 2007. *Contentious Politics.* Oxford: Oxford University Press.

Torche, F. 2011. "Is a college degree still the great equalizer? Intergenerational mobility across levels of schooling in the United States." *American Journal of Sociology* 117(3): 763–807.

Tsoukas, H. 1996. "The firm as a distributed knowledge system: A constructionist approach." *Strategic Management Journal* 17: 11–25.

———. 2005. "Do we really understand tacit knowledge?" In *Managing Knowledge: An Essential Reader*, 107–126. Edited by Stephen Little and Tim Ray. London: Sage.

———. 2009. "A dialogical approach to the creation of new knowledge in organizations." *Organization Science* 20(6): 941–957.

Tsoukas, H., and N. Mylonopoulos. 2004. "Introduction: Knowledge construction and creation in organizations." *British Journal of Management* 15(S1): S1–S8.

Tsoukas, H., and E. Vladimirou. 2001. "What is organizational knowledge?" *Journal of Management Studies* 38: 973–993.

Turner, J. R., and A. Keegan. 1999. "The versatile project-based organization: Governance and operational control." *European Management Journal* 17(3): 296–309.

———. 2001. "Mechanisms of governance in the project-based organization: Roles of the broker and steward." *European Management Journal* 19(3): 254–267.

Turner, R., and L. Killian. 1972. *Collective Behavior.* Englewood Cliffs, N.J.: Prentice-Hall.

Turner, V. W. 1982. *From Ritual to Theatre: The Human Seriousness of Play.* New York: Performing Arts Journal Publications.

Tushman, M. L., and P. Anderson. 1986. "Technological discontinuities and organizational environments." *Administrative Science Quarterly* 31: 439–465.

Tushman, M. L., and C. A. O'Reilly III. 1996. "Managing evolutionary and revolutionary change." *California Management Review* 38(4): 8–28.

Uzzi, B. 1997. "Social structure and competition in interfirm networks: The paradox of embeddedness." *Administrative Science Quarterly* 42(1): 35–67.

Van de Ven, A. H., and T. J. Hargrave. 2004. *Social, Technical, and Institutional Change.* Oxford: Oxford University Press.

Van der Valk, W., F. Wynstra, and B. Axelsson. 2009. "Effective buyer-supplier interaction patterns in ongoing service exchange." *International Journal of Operations and Production Management* 29(8): 807–833.

Vanderstraeten, R. 2002. "Parsons, Luhmann and the theorem of double contingency." *Journal of Classical Sociology* 2(1): 77–92.

Venkatesh, S. 2013. "Resilience and rebuilding for low-income communities: Research to inform policy and practice." Keynote remarks, Federal Reserve Bank of Atlanta, Washington, D.C., April 11, https://frbatlanta.org/news/conferences/2013/130411-resilience-rebuilding/media/venkatesh-transcript.aspx.

Vissa, B. 2012. "Agency in action: Entrepreneurs' networking style and initiation of economic exchange." *Organization Science* 23(2): 492–510.

Von Hippel, E. 1994. "'Sticky information' and the locus of problem solving: Implications for innovation." *Management Science* 40(4): 429–439.

Weick, K. 1969. *The Social Psychology of Organizing*. Reading, Mass.: Addison-Wesley.

———. 1979. *The Social Psychology of Organizing*. 2nd ed. New York: McGraw-Hill.

———. 1995. *Sensemaking in Organizations*. Vol. 3. Thousand Oaks, Calif.: Sage.

———. 1998. "Introductory essay: Improvisation as a mindset for organizational analysis." *Organization Science* 9(5): 543–555.

Weick, K. E., and R. E. Quinn. 1999. "Organizational change and development." *Annual Review of Psychology* 50(1): 361–386.

Weick, K. E., and K. H. Roberts. 1993. "Collective mind in organizations: Heedful interrelating on flight decks." *Administrative Science Quarterly* 38: 357–381.

Weick, K. E., and K. M. Sutcliffe. 2006. "Mindfulness and the quality of organizational attention." *Organization Science* 17(4): 514–524.

Weick, K., K. Sutcliffe, and D. Obstfeld. 1999. "Organizing for high reliability: Processes of collective mindfulness." In *Research in Organizational Behavior*. Vol. 21. Edited by B. Staw and R. Sutton. Stamford, Conn.: JAI Press.

———. 2005. "Organizing and the process of sensemaking." *Organization Science* 16(4): 409–421.

Wertsch, J. V. 1991. "The problem of meaning in a sociocultural approach to mind." In *Toward the Practice of Theory-based Instruction: Current Cognitive Theories and Their Educational Promise*, 31–46. Edited by A. McKeough and J. L. Lupart. Hillsdale, N.J.: Lawrence Erlbaum Associates.

White, H. C. 1970. *Chains of Opportunity: System Models of Mobility in Organizations*. Cambridge, Mass.: Harvard University Press.

———. 2008. *Identity and Control: How Social Formations Emerge*. Princeton, N.J.: Princeton University Press.

Winograd, T., and F. Flores. 1986. *Understanding Computers and Cognition: A New Foundation for Design*. Norwood, N.J.: Ablex.

Winter, S. G. 1987. "Knowledge and competence as strategic assets." In *The Competitive Challenge: Strategies for Industrial Innovation and Renewal*, 159–184. Edited by D. J. Teece. Cambridge, Mass.: Ballinger.

———. 1994. "Organizing for continuous improvement: Evolutionary theory meets the quality revolution." In *Evolutionary Dynamics of Organizations*, 90–108. Edited by J. Baum and J. Singh. New York: Oxford University Press.

———. 2003. "Understanding dynamic capabilities." *Strategic Management Journal* 24(10): 991–995.

———. 2011. "Problems at the foundation? Comments on Felin and Foss." *Journal of Institutional Economics* 7(2): 257–277.

———. 2013. "Habit, deliberation, and action: Strengthening the microfoundations of routines and capabilities." *Academy of Management Perspectives* 27(2): 120–137.

Wood, W., J. M. Quinn, and D. Kashy. 2002. "Habits in everyday life: Thought, emotion, and action." *Journal of Personality and Social Psychology* 83(6): 1281.

Xiao, Z., and A. S. Tsui. 2007. "When brokers may not work: The cultural contingency of social capital in Chinese high-tech firms." *Administrative Science Quarterly* 52(1): 1–31.

Zander, U., and B. Kogut. 1995. "Knowledge and the speed of the transfer and imitation of organizational capabilities: An empirical test." *Organization Science* 6(1): 76–92.

Zollo, M., and S. G. Winter. 2002. "Deliberate learning and the evolution of dynamic capabilities." *Organization Science* 13(3): 339–351.

Index